Cell Biology Monographs
Continuation of Protoplasmatologia

Founded by

L. V. Heilbrunn, Philadelphia, Pa., and F. Weber, Graz

Vol. 1

Springer-Verlag
Wien New York

The Lytic Compartment of Plant Cells

Ph. Matile

Springer-Verlag

Wien New York

Prof. Dr. PHILIPPE MATILE

Department of General Botany,
Swiss Federal Institute of Technology Zurich (ETHZ),
Switzerland

© 1975 by Springer-Verlag/Wien
Printed in Austria by Adolf Holzhausens Nfg., Wien

With 59 Figures and 40 Plates

Library of Congress Cataloging in Publication Data. Matile, Philippe, 1932—.The lytic compartment of plant
cells. (Cell biology monographs; v. 1.) Bibliography: p. 1. Plant cells and tissues. 2. Lysosomes. I. Title. II. Series.
QK725.M34.581.8'761.75-5931

ISBN 3-211-81296-2 Springer-Verlag Wien-New York
ISBN 0-387-81296-2 Springer-Verlag New York-Wien

Dedicated to Prof. A. Frey-Wyssling,
a pioneer in the elucidation of relationships
between cell structure and metabolism,
on his 75th birthday

About twenty years ago, Prof. FRIEDL WEBER (Graz University) and Prof. L. V. HEILBRUNN (University of Pennsylvania) conceived the idea for the handbook "Protoplasmatologia" at a time when the state of knowledge in the field of cell biology still permitted one to think of an all-encompassing handbook in the classical sense. Since 1953 fifty-three volumes with a total of about 9,500 pages have been published. The very rapid developments in this area of science, especially during the last decade, have led to new insights which necessitated some alterations in the original plan of the handbook; also, changes in the board of editors since the death of the founders have brought about a reorientation of viewpoints.

The editors, in agreement with the publisher, have decided to abandon the confining limits of the original disposition of the handbook altogether and to continue this work, in a form more appropriate to current needs, as an open series of monographs dealing with present-day problems and findings in cell biology. This will make it possible to treat the most modern and interesting aspects of the field as they arise in the course of contemporary research. The highest scientific, editorial and publishing standards will continue to be maintained.

Editors and publisher

Preface

Almost twenty years ago DE DUVE discovered the existence in rat liver cells of a novel class of subcellular structures which he termed lysosomes. As the lysosomes are the seat of various hydrolytic enzymes this discovery greatly stimulated the interest in cellular digestive processes. It is now recognized that compartmentation of hydrolases is an outstanding example of the dependence of metabolic functions on cell structure. The important role of lytic processes in plant metabolism made it interesting to correlate the facts on the various phenomena and present them from the point of view of cellular compartmentation. It is not intended to survey *in extenso* the literature on the various aspects of localization and function of hydrolases in plant cells as this would be an impossible task. The aim of this monograph is to emphasize the significance of hydrolase compartmentation rather than to strive for complete coverage of work on hydrolases in which other viewpoints have been considered.

I thank Miss SONIA TURLER for her great help with the manuscript. I also wish to acknowledge the assistance given by Dr. ELSA HÄUSERMANN, Mrs. DORLI FURRER, and Mrs. SILVIA STÜNZI. Moreover, I am indebted to my collaborators and particularly to my colleague Dr. A. WIEMKEN for frequent and stimulating discussions on the subject.

Zurich (Switzerland), February 1975   PH. MATILE

Contents

Abbreviations

AG   Aleurone grain
CW   Cell wall
D   Dictyosome
DV   Dictyosome vesicle
DNA   Deoxyribonucleic acid
DNase Deoxyribonuclease
ER   Endoplasmic reticulum
M   Mitochondrion
N   Nucleus
P   Plastid
PL   Plasmalemma
R   Ribosome
RER   Rough surfaced endoplasmic reticulum
RNA   Ribonucleic acid
RNase Ribonuclease
S   Spherosome
SER   Smooth surfaced endoplasmic reticulum
T   Tonoplast
V   Vacuole

Introduction: Lysis and the Lytic Compartment

The metabolism of natural ecosystems is characterized by a balance between the primary production in the photoautotrophs and the utilization and eventual mineralization of primary products along the complex food chains. Obviously, the lytic phenomena such as degradation of macromolecular compounds produced in plants by herbivorous animals and in saprophytic microorganisms play a prominent role. It is often less appreciated that the plants themselves produce lytic enzymes that are capable of degrading their own cellular constituents. Hence, lysis must have a significant function in the plant's own metabolism.

Let us briefly consider a few important and obvious manifestations of lysis in plants. First of all, plants and their organs and even individual cells have a limited and determined life span: they grow, become mature, senesce and eventually die off. The analysis of abscised leaves demonstrates that a considerable proportion of the cytoplasmic substances, especially macromolecules containing nitrogen are broken down in the course of ageing. Another developmental process which is characterized by conspicuous lytic events is germination of seeds the macromolecular reserves of which are mobilized upon the induction of growth in the embryo.

Whilst senescence and germination appear to concern the lysis of cytoplasmic constituents, other phenomena clearly point to the lysis of cell wall material. Abscission of leaves, accomplished by the degradation of cell wall polysaccharides in the abscission zone is an example of this. The destruction of wood by saprophytic fungi or the invasion of plant cells by parasitic fungi through the dissolution of the cell walls of the host are further indications of extracytoplasmic lysis.

A superficial consideration of lytic phenomena in plants may lead to the opinion that synthesis and breakdown are always separated in time. Indeed, accumulation and mobilization of reserves are separated by seed dormancy which may last for several hundred years as demonstrated by the famous example of *Nelumbo nucifera*. In the development of leaves, the synthesis of cytoplasmic macromolecules in the growing organ again seems to be temporally separated from their decline upon senescence. In actual fact, the two polar processes, anabolic and catabolic, also occur simultaneously. This was recognized as early as in 1897 by the German plant physiologist PFEFFER who anticipated what is now known as cellular turnover of protein taking place in developing plant organs: „Voraussichtlich schreitet die Zerspaltung von Proteinstoffen ununterbrochen fort und kommt wahrscheinlich auch nicht zum Stillstand, wenn die ausgewachsene Pflanze nur noch für die Erhaltung und den Betrieb des Bestehenden zu sorgen hat." Proof of the existence of protein turnover was not feasible experimentally in PFEFFER's days and even now it is difficult to evaluate turnover rates quantitatively. However, it can be taken for granted that cellular macromolecules essential for life are metabolically labile, that is, they are synthesized and broken down simultaneously in individual cells. This appears to be true even for constituents of the plant cell wall.

The significance of metabolic lability is not always obvious. On the one hand it appears to be a key to the understanding of such fundamental processes as differentiation and development which imply the remodeling of structures and of the biochemical machinery through lysis of useless cell constituents and synthesis of those required. Also, turnover reactions seem to be a basis for adaptation of metabolism to the ever changing environmental conditions.

From a biochemical point of view the lytic processes are much less attractive than is the synthesis of macromolecular cell constituents. Molecular biologists have elucidated a most complex organization and control of the synthesis of such complicated biopolymers as proteins and nucleic acids. Nuclei with their highly structured chromosomes and nucleoli, polysomes and cellular membrane systems participate in the transcription and translation of the genetic material. In sharp contrast to this elaborate biochemical web the breakdown reactions are nothing but hydrolysis catalyzed by a number of mostly unspecific enzymes. This has the following important consequence.

Cellular turnover implies that the polar processes, anabolic and catabolic, must be separated in space. In other words, turnover requires compartmentation of processes which otherwise would interfere with each other. The unspecific nature of hydrolases such as proteinases and nucleases requires that the cells must keep them separated from the proteins and nucleic acid involved in anabolism. It is now clear that such a membrane bounded *lytic compartment of plant cells* exists and a variety of lytic phenomena can be associated with its operation. This compartment has many features in common with its analogue in animal cells; corresponding reference to this will be made occasionally.

Summarizing these introductory remarks it should be pointed out that the polarity of processes going on in metabolism requires a duality of views and experimental approaches: dealing with lysis implies that the *biochemical aspects* of degradation must always be complemented by the *morphological aspects* of compartmentation. Incidently, the importance of compartmentation of hydrolases appears in the work of biochemists dealing with nucleic acid synthesis and metabolism: When preparing biologically active nucleic acids the careful removal or inactivation of nucleases is nothing less than separating what is located in different compartments of the living cell.

1. Hydrolases and their Localization

1.1. Classification and Properties of Hydrolases

As this monograph deals with the cellular location and metabolic function of hydrolytic enzymes, a brief summary of some important properties of this group of biocatalysts is justified. For more detailed information the reader is referred to the standard treatises on enzyme biochemistry.

Hydrolases can be classified according to the type of chemical bond cleaved by them. The main groups of hydrolases are thus, *peptidases* hydrolyzing peptide bonds, *esterases* attacking various ester bonds of phosphoric acid, sulfuric acid and carbonic acids, and *glycosidases* responsible for the cleavage of glycosidic bonds between sugar residues.

1.1.1. Peptidases

A wide variety of peptidases from higher plant tissues (review by RYAN 1973) as well as from fungi have been detected, isolated and characterized. They fall into two categories according to their operation on peptides: *endopeptidases* split up large polypeptide chains into smaller fragments by cleaving internal peptide bonds of polypeptides, *exopeptidases* hydrolyze the terminal amino acid residues. The latter act either specifically on amino acids at the amino-terminal of peptides (aminopeptidases) or at the carboxy-terminal (carboxypeptidases).

Classification of peptidases can also be based on their catalytic mechanism. Four classes have been distinguished: proteinases whose active center contains serine residues (serineproteinases) or sulfhydryl groups (sulfhydryl-proteinases), metallo-peptidases whose activity depends on the presence of cations such as Co^{++} or Zn^{++} and, finally, acid proteinases which are optimally active in an acidic environment.

Examples of all of these types have been encountered in plants, and in certain objects that have been studied in detail the presence of various types of peptidases has been recognized. An example illustrating the presence of a complete proteolytic machinery in a single cell is given by the yeast *Saccharomyces cerevisiae*. The elucidation of this machinery is mainly the merit of LENNEY, HAYASHI and their collaborators (LENNEY 1956, LENNEY and DALBEC 1967, 1969, HATA et al. 1967 a, b, DOI et al. 1967, HAYASHI et al. 1969, 1973, HAYASHI and HATA 1972). In extracts from yeast two endopeptidases can be distinguished. Proteinase A is an acid proteinase which hydrolyzes denatured proteins optimally at a pH around 3; it has similar properties as pepsin. Proteinase B is active towards denatured casein at pH 9 and resembles trypsin in that it exhibits esterolytic activity. It is a rather labile sulfhydryl-proteinase which in extracts is partially present in the inactive form as a complex of enzyme and inhibitor-protein. Protease C, a serine enzyme, has a unique carboxypeptidase activity and also possesses a specific inhibitor-protein, which may be destroyed in the presence of proteinase A. Recently, LENNEY has shown that a specific inhibitor-protein exists also for proteinase B; like the inhibitor of A this inhibitor-protein is heat stable.

Hence, the yeast cell possesses a set of three proteases and a corresponding set of three specific inhibitor-proteins. In addition, the occurrence of several aminopeptidases has been demonstrated (MATILE *et al.* 1971).

In thoroughly investigated tissues such as storage organs of seeds the existence of similar sets of proteolytic enzymes has been established. Moreover, proteinase inhibitors are widely distributed in the plant kingdom; as demonstrated by the almost classical example of the soybean trypsin inhibitor, these proteins interfere with animal proteinases whereas inhibitor-proteins present in animal tissues seem to have no effect on plant proteinases (see RYAN 1973).

1.1.2. Esterases

1.1.2.1. Phosphomonoesterases

Perhaps the most famous among plant hydrolases are the phosphate-esterases which unspecifically cleave phosphomonoesters. They are conventionally assayed using β-glycerophosphate or p-nitrophenylphosphate as substrates. The popularity of these *unspecific phosphatases* is due to the convenience of assessment and also to the availability of safe cytochemical techniques for localizing the activity in cells and tissues (GOMORI-reaction). In general, higher plants seem to contain only unspecific acid phosphatases; fungi produce both acid and alkaline enzymes.

An unspecific acid phosphatase purified from tobacco leaves splits a wide variety of phosphate esters including sugar phosphates, nucleotides and phosphoenolpyruvate; even anhydrous bonds of phosphoric acid in ATP are hydrolyzed by this enzyme (SHAW 1966). Considering the importance of phosphate esters among metabolic intermediates, the striking unspecificity of phosphomonoesterase implies the necessity of compartmentation.

Surveys of phosphatases carried out by gel filtration or disc-electrophoresis have revealed the presence of a number of distinct molecular forms (isoenzymes) of phosphomonoesterase differing in their relative substrate specificity, some being more active on substrates such as β-glycerophosphate, others preferentially hydrolyzing nucleotides (*e.g.*, ROBERTS 1970, MEYER *et al.* 1971).

Acid phosphatases have also been purified from fungi; their unspecificity with regard to the substrates attacked appears, for example, from a study of KUO and BLUMENTHAL (1961) on a phosphomonoesterase extracted from mycelia of *Neurospora crassa*. An unspecific alkaline phosphatase present in this mould is optimally active at pH 9.0–9.5; in contrast to the acid enzyme its stability depends on the presence of metal ions (NYC *et al.* 1966).

In addition to these unspecific phosphomonoesterases, plant cells may contain specific phosphatases such as phytase which hydrolyzes myoinositolhexa-phosphate, a major storage form of phosphate in seeds.

1.1.2.2. Phosphodiesterases

In contrast to phosphomonoesterases, the diesterases cleave ester bonds of phosphate linking two sugar residues. Dinucleotides would be typical substrates of these enzymes. The most prominent among phosphodiesterases are the *nucleases*. Both, ribonucleases (RNases) and deoxyribonucleases (DNase) are present in cells of higher plants, as also in those of fungi and other micro-organisms. Much more is known about plant RNases (review by DOVE 1973) than about DNase.

Specific RNases, which attack RNA in an endo-fashion and have a relative purine specificity have been characterized. *In vitro* these will liberate purine -3′-nucleotides from cyclic nucleotides. In addition, endonucleases producing 5′-nucleotides from both RNA and DNA and hydrolases which specifically attack native DNA have been described. It appears upon gel electrophoresis that several molecular forms of RNase are usually present in plant tissues.

1.1.2.3. Acyl-Esterases

Another important type of esterases are those which cleave esters of carboxylic acids. *Unspecific esterases* of this group hydrolyze substrates such as p-nitrophenylesters or β-naphthylesters of acetic acid. They seem to be relatively specific in attacking esters of short chain carbonic acids in comparison with esters of long chain fatty acids (*e.g.*, NORGAARD and MONTGOMERY 1968); several distinct isoenzymes can be distinguished either by recording inhibitor responses or upon electrophoretical separation (SAHULKA and BENEŠ 1969). These carboxylic esterases are ubiquitous in the plant kingdom.

Unspecific esterases seem to be inactive, however, on triglycerides. These important reserve substances are hydrolyzed by *lipases;* acid lipase of castor bean, a complex of apoenzyme, lipid cofactor, and a protein activator is an example of this type of specific esterase (ORY 1969). Similar enzymes have been detected in storage organs of other oleaginous seeds. Triglyceride-lipases are also produced by a variety of fungi. A further group of esterases is represented by phospholipases which specifically split fatty acids from phospholipids and glycolipids. Another example of a specific esterase is chlorophyllase which catalyzes the saponification of chlorophylls into phytol and chlorophyllides. The natural substrates of arysulfatases, the presence of which has been established in fungi (see SCOTT and METZENBERG 1970) and in certain higher plants (see BAUM and DODGSON 1967) are unknown. In higher plants the sulfogalactolipids of plastids or sulfo-derivatives of phenolic compounds formed by certain species represent potential substrates of this type of esterase.

1.1.3. Glycosidases

This brief account on glycosidases starts with a survey of plant cell constituents which represent potential substrates. These comprise a variety of structural polysaccharides (cellulose, β-1,3-, β-1,6-, and α-1,3-glucans, hemicelluloses, pectic substances, mannans, chitin etc.), the reserve polysaccharides (starch in its two forms of the unbranched α-1,4-glucan amylose and the branched α-1,4- and α-1,6-amylopectin, glycogen, the linear β-1,3-glucan paramylon, the fructosan inulin etc.), the various di- and oligosaccharides and, finally, a large number of glycosides, for example, of phenolic substances such as anthocyanins. A corresponding variety of glycosidases has been detected in the plant kingdom. The degradation of polysaccharides containing several different sugar residues (hemicelluloses) or different glycosidic linkages (*e.g.*, α-1,4- and α-1,6-linkages of glucose in amylopectin; β-1,3- and β-1,6 of glucose in yeast glucan) requires the action of the appropriate glycosidases. In addition, the degradation is usually accomplished by the joint action of endo- and exoglycosidases. The saccharification of amylopectin involves α-amylase, which breaks down α-1,4-polyglucose chains to short dextrin segments, β-amylase, fragmenting these into maltose residues, β-glucosidase responsible for the hydrolysis of maltose and amylopectin-1,6-glycosidase which is the debranching enzyme.

Glycosidases which attack polysaccharides in an endo-fashion can be assayed by following the decrease of viscosity in the substrate solution. Endopolygalacturonases (pectinase), endo-β-1,3-glucanases, and cellulases are preferentially measured viscosimetrically whereby the water-soluble carboxymethyl derivatives of water-insoluble glucans are used as substrates. A convenient method of assessing the depolymerization of starch by the action of α-amylases is based on the decrease in intensity of the starch-iodine complex. In enzyme preparations containing endo- and exoglycosidases it is difficult to distinguish between these enzymes by measuring the liberation of sugars in a test for reducing groups. Both, endo- and exoglycosidases liberate reducing groups and this is why chromophoric substrates are preferably used for the estimation of exoglycosidases. Substrates like p-nitrophenyl-glycosides determine the type of glycosidic bond as well as the sugar residue with which the enzyme binds. A disadvantage of these convenient substrates is that the connection of the tested activity with the natural substrate of the enzyme is lost. A β-glucosidase purified from petals of *Impatiens balsamina* showed a requirement for an aromatic aglycone and a single β-linked glycoside; it was active towards p-nitrophenyl-β-glucopyranoside but inactive towards the disaccharide cellobiose (BOYLEN and HAGEN 1969). Hence, the study of functional aspects of glycosidases requires work with the natural substrates.

Plant glycosidases are optimally active at slightly acidic pH values. Comparatively little information about other requirements for the catalytic action is available and this may simply reflect the independence of most glycosidases from a specific milieu.

As in the case of peptidases and esterases the occurrence of multiple forms of glycosidases in a single tissue has been reported: *e.g.*, in the endosperm of cereals a number of amylase-isoenzymes are present (see SCANDALIOS 1969).

1.1.4. General Remarks

Most plant hydrolases are *glycoproteins*. There may be exceptions and in many cases where hydrolases have been purified, this particular aspect has not been considered. In addition, the carbohydrate content may be small enough to escape analysis. The following few examples demonstrate the differences of degrees of glycosylation. Baker's yeast invertase contains 50% carbohydrate; about 30 chains of mannan are present per molecule of enzyme and appear to be linked to the polypeptide *via* glucosyl-aminyl-asparagine bonds (NEUMANN and LAMPEN 1969). Enzymically active forms of yeast invertase with carbohydrate contents as low as 10% have been isolated (WAHEED and SHALL 1971); hence, the degree of glycosylation appears to vary greatly in a distinct hydrolase. In an acid phosphatase produced by *Neurospora crassa* mannose and glucosamine account for 9.5% of the molecular weight (JACOBS *et al.* 1971) and, to conclude with an example of a higher plant hydrolase, purified α-amylase from barley endosperm contains only some glucosamine (VARNER, personal communication).

The significance of glycosylation of hydrolases is not yet clear. It is possible that the carbohydrate moieties contribute to the stability of these enzymes under the conditions of the lytic compartment. Since they may be exposed to proteinases present in the same compartment, the polysaccharide could have a function in protecting the hydrolases against proteolytic inactivation.

It is important to know that inhibitors present in crude homogenates from plant tissues may falsify the determination of hydrolase activities. Such inhibitors may give the misleading impression of the existence of zymogen-forms of hydrolases which require some sort of proteolysis for their activation. The example of an apparent zymogenic form of protease C in yeast which is in fact an enzyme-inhibitor protein-complex has already been discussed. An amylopectin-1,6-glucosidase present in an inactive form in pea seeds requires a limited proteolysis for activation (MAYER and SHAIN 1968); the zymogenic character of this enzyme could also be apparent, the proteolytic activation merely being caused by the destruction of an inhibitor protein.

1.2. External and Internal Locations of Hydrolases

Multiple locations in plant cells are typical for many hydrolases. Aryl sulfatase present in conidia of *Neurospora crassa* provides an illustrative example of this phenomenon. Little is known about the metabolic function of this particular hydrolase, yet SCOTT and METZENBERG (1970) presented a convincing biochemical analysis of its locations. Using p-nitrophenyl sulfate as an artificial substrate, a fraction of the total sulfatase activity (which appears upon complete rupture of the spores) can be assessed in intact conidia. This substrate-accessible enzyme must be located outside the permeability barrier for the substrate, which is presumably the plasma membrane. If conidia are suspended in buffer a substantial proportion of this external activity is removed; further washings with buffer and even treatments with KCl of high ionic strength or EDTA fail to remove additional external sulfatase. Hence, some external enzyme is located on the cell surface in a free form, whilst some activity is cell-bound. The latter appears to be tightly bound to the cell wall; it is *mural*.

The external, *patent* enzyme can be detected in suspensions of intact *Neurospora* conidia, but additional cell-bound activity is only detectable if the spores are first treated with agents known to make membranes permeable. This additional enzyme is *cryptic;* it becomes accessible to the substrate by the exposure of spores to organic solvents or to the polyenic antibiotic nystatin. The patent sulfatase activity can be inactivated by brief treatments of intact conidia with weak hydrochloric acid without loss of conidial viability. The cryptic activity is acid resistant unless the spores have been made permeable. All of these phenomena and additional experimental data (SCOTT and METZENBERG 1970) support the triple locations of aryl-sulfatase in *Neurospora* conidia sketched in Fig. 1. Obviously, it is not possible on the basis of these data to distinguish whether the plasmalemma alone, or together with additional membrane envelopes of cytoplasmic organelles containing

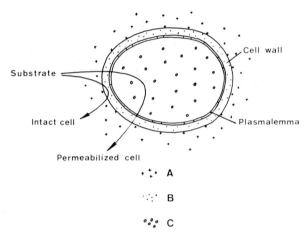

Fig. 1. Multiple location of aryl-sulfatase in conidia of *Neurospora crassa*. Drawn according to data presented by SCOTT and METZENBERG (1970). *A* extracellular, released enzyme; *B* cell-bound, patent (mural) enzyme; *C* cell-bound, cryptic enzyme. *A* and *B* represent external, secreted enzyme, *C* is internal hydrolase.

sulfatase, is responsible for the cryptic nature of the enzyme (see Plate 4 *D*). According to SCOTT *et al.* (1971) the cryptic enzyme is, in fact, located within a membrane bound cytoplasmic particle. Subsequently the terminology summarized in Fig. 1 will be used in describing locations of hydrolases with respect to cells.

1.3. Extracellular Release of Hydrolases

The release of hydrolases into the environment is a common feature in fungi. Since this phenomenon has an important technological aspect much more is known about the biochemical properties of hydrolases that can be isolated from culture filtrates and used in industrial food processing than about the biological aspects of hydrolase release. Hydrolase release is also important from the point of view of biological experimentation since poly-saccharidases released from fungi and other microorganisms are widely used for preparing cell wall-free protoplasts, so called spheroplasts.

It is evident that hydrolase release in plant cells can be compared with a metabolic arm which seizes substrates present in the environment. The

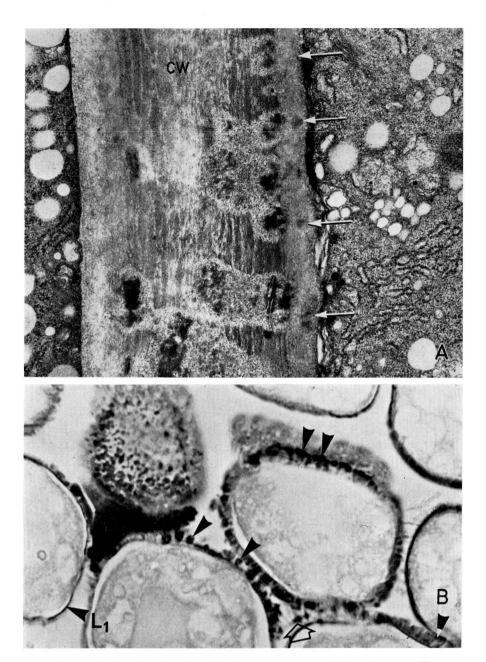

Plate 1. Perforation of walls and hydrolase secretion in barley aleurone cells.

A: Cross sectional view of aleurone cells induced to secrete hydrolases. Near the basal pole of the cell on the right disorganization of cell wall material is extensive and extends from the plasmalemma via narrow channels (arrows) to large disarrayed regions further in the wall. ×21,000. Courtesy of E. L. Vigil. *B:* Cytochemical localization of acid phosphatase in barley aleurone cells induced to secrete hydrolases. The reaction product appears in the digested regions of the cell walls (arrows). L_1 marks enzyme which has not penetrated into an undigested portion of the cell wall. ×1,800. Courtesy of E. Ashford and J. V. Jacobsen.

nutritional aspect of this is obvious. Fungi which are able to grow on macro-molecular substrates such as proteins, starch, cellulose, chitin or nucleic acids release the corresponding hydrolases. The phenomenon of extracellular hydrolase release is, however, not restricted to fungi. It occurs in higher plants as well. A famous example is represented by insectivorous plants; for instance the pitchers of *Nepenthes,* a tropical genus, produce a liquid which contains several proteinases (*e.g.,* AMAGASE 1972). It seems logical that a fungal mycelium growing on timber or on an insect cuticle has no other choice than to digest the comparatively gigantic lumps of food extracellularly. Therefore, fungi continue to release hydrolases even if they are cultured in the laboratory on macromolecules small enough to diffuse through the cell wall. Plant cells seem to be unable to take up large particles into the cytoplasm and this appears to be due to the existence of a cell wall with pores of only submicroscopic dimensions. Yet the extracellular release of hydrolases demonstrates that, conversely, macromolecules can move through plant cell walls. Hence, one important prerequisite of extra-cellular hydrolase release is the existence of wall pores with diameters larger than those of the hydrolases. The necessity of adequate wall porosity is nicely illustrated by the aleurone cells of barley endosperm. The cells are responsible for the mobilization of reserves stored in the adjacent starchy endosperm. Upon seed germination this is achieved by the extra-cellular release of various hydrolases such as α-amylase, proteinase and others. However, the secretory aleurone cells have thick cell walls which, appear to be impenetrable for large molecules. TAIZ and JONES (1970) observed that secretion of α-amylase is preceded by the release of β-*1,3-glucanase,* which evidently perforates the cell walls of the aleurone layer by degrading the corresponding polysaccharides (Plate 1 *A*). Using acid phos-phatase activity (another hydrolase released from this tissue) for following the pathway of enzyme release cytochemically, ASHFORD and JACOBSEN (1974) were able to show that the wall channels produced by glucanase act as preferential routes for the acid phosphatase which is released from the aleurone layers (Plate 1 *B*). Indeed, no release of acid phosphatase into the bathing medium occurred, if the tissue had not been induced to secrete β-glucanase.

1.4. Hydrolases Associated with Cell Walls

1.4.1. Demonstration and Estimation of Hydrolases Associated with the Cell Wall

Cytochemical techniques seem to be ideal for the elucidation of hydrolase compartmentation; they provide the dual information required about the enzyme *and* its presence or absence in distinct cell spaces. Indeed, cyto-chemists have accumulated a vast amount of evidence for the presence of hydrolases in plant cell walls.

The levels of mural β-glycerophosphatase appear to vary in the different tissues of a plant organ. In barley root tips acid phosphatase is mainly present in the walls of the epidermal and subepidermal cells (Plate 2 *A*, HALL 1969).

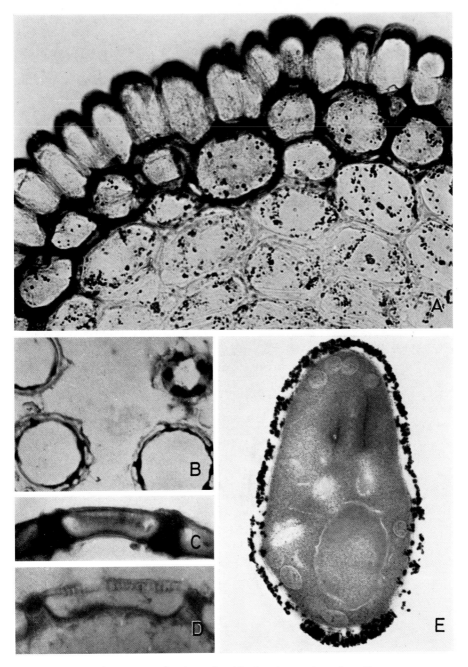

Plate 2. Localization of acid phosphatase in cell walls.

A: β-Glycerophosphatase in the cell walls of epidermal and subepidermal cells of barley root tip. ×615. Courtesy of J. L. Hall. *B:* Freeze-sectioned young spores of *Malvaviscus arboreus* showing concentration of acid phosphatase in the poral intine; ×250; in the mature pollen wall the enzyme is concentrated in the pores (*C* ×1,000); *D:* Control without substrate; ×1,000. *B–D:* Naphtyl-phosphate-pararosanilin reaction. Courtesy of J. Heslop-Harrison. *E:* Reaction product of acid phosphatase in the cell wall of *Saccharomyces cerevisiae* cultured in a low-phosphate medium. Ultrathin frozen section; poststained with phosphotungstic acid. ×11,000. Courtesy of H. Bauer.

Similar cytochemical observations on hydrolase localizations in specific tissues have been reported repeatedly (*e.g.*, McLean and Gahan 1970, Hébant 1973). Even in an individual cell, phosphatase may be concentrated in certain regions of the wall as illustrated by its predominant presence in the poral intine of apertured pollen grains (Plate 2 *B–D*, Knox and Heslop-Harrison 1970). The distribution of reaction product of leucineaminopeptidase in yeast cells (Plate 4 *C*) suggests that this enzyme is localized in the cell wall (Reiss 1972). The existence of mural phosphatase appears also from numerous localization studies made on the electron microscopy level (*e.g.*, Halperin 1969, Knox and Heslop-Harrison 1971, Catesson *et al.* 1971, Günther *et al.* 1967, Bauer and Sigarlakie 1973, Poux 1970, Sexton *et al.* 1971, Maier and Maier 1972), as shown in the micrographs of Plate 2 *E* and Plate 3.

Apart from technical difficulties inherent in enzyme cytochemistry (fixation, immobilization of reaction products, conditions of incubation, controls) the method is restricted to comparatively few hydrolase activities, especially in electron microscopy which requires sufficient electron opacity of the reaction product. Hydrolases ultimately yielding phosphate, that is, various phosphatases or nucleases (the nucleotides produced being hydrolyzed to inorganic phosphate by purified phosphatase added to the incubation mixture) have mostly been localized electron microscopically in a Gomori-type of reaction. The choice is wider for light microscopical purposes since azo-coupling methods in conjunction with naphthyl-substrates, *e.g.*, glucosides, can be used (*e.g.*, Ashford and McCully 1970 a). The osmication of insoluble azo-dyes produced may bring this method into the range of electron microscopy (Knox and Heslop-Harrison 1971). Using azo-coupling reagent containing a heavy metal Gahan and McLean (1969) have been able to localize naphthylesterase activity in the cell wall.

A promising approach to hydrolase localization is provided by immunocytochemistry. The advantage of this method is that a specific enzyme protein is stained by its antibody coupled directly or indirectly with a fluorochrome (see Weston and Poole 1973), whereas in classical enzyme cytochemistry merely the activity is localized. The necessity of extensive purification of the enzyme to be localized immunocytochemically is, of course, a disadvantage. Examples of successful immunofluorescent demonstration of mural invertase in *Neurospora* (Chang and Trevithick 1970, 1972 a) and in *Saccharomyces* (Tkacz and Lampen 1973) have been presented.

Still another shortcoming of cytochemical techniques is the lack of quantitative information due to the difficulties of quantifying the hydrolase localizations observed. Complementary biochemical approaches are therefore necessary. One of these has already been touched on above: the assessment of mural hydrolase activities using chromophoric substrates that diffuse only very slowly or not at all through the plasma membrane of intact cells (Scott and Metzenberg 1970). In addition to acid phosphatase (Ridge and Rovira 1971, Matile 1973) β-glucosidase, β-galactosidase and other glycosidases (Stenlid 1957, Matile 1973, Keegstra and Albersheim 1970, Klis 1971) are accessible for p-nitrophenyl-substrates in intact tissues. The hydrolases

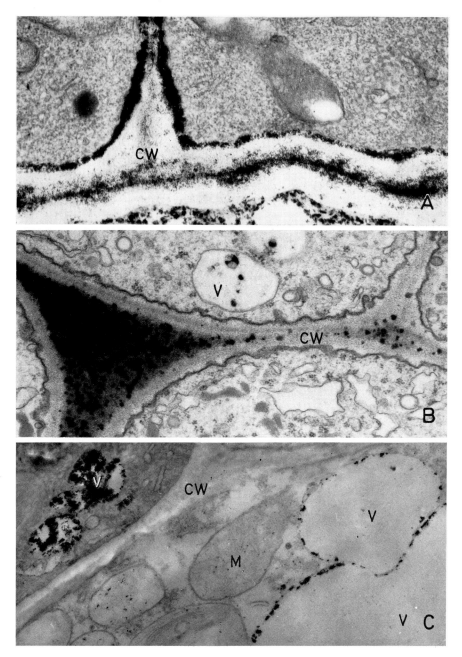

Plate 3. Cytochemical localizations of acid phosphatase.

A: Product of Gomori-reaction in cell walls of meristematic root cells of *Cucumis sativus.* ×30,000. Courtesy of N. Poux. *B:* Localization in cell walls of cultured carrot cells. ×32,000. Courtesy of W. Halperin. *C:* Absence of acid phosphatase from cell walls in the petiolary nectarium of *Vicia faba.* Note the presence of acid phosphatase in the vacuoles. ×23,000. Courtesy of J. Figier.

seem to be mural since the activity is not released from the tissue (Fig. 2). Patent cell-bound p-nitrophenyl β-glucosidase has also been discovered in *Neurospora* conidia and young mycelia (EBENHART and BECK 1970) and aminopeptidase in the conidia of this fungus (MATILE 1965). In another technique for demonstrating mural hydrolase activities natural substrates are used, the uptake and metabolism of reaction products being prevented in the presence of appropriate inhibitors. By blocking the glucose transport in *Mucor rouxii* in the presence of UO_2^+, FLORES-CARREON *et al.* (1970) were able to

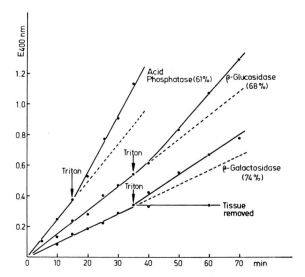

Fig. 2. External and cryptic activities of hydrolases in intact segments of corn root tips. The activities toward p-nitrophenyl substrates were estimated both in the absence and in the presence of 1% Triton-x-100. The figures in parenthesis indicate the apparent external (mural) activities as percentages of total activities in the permeabilized tissue. From MATILE (1973).

estimate cell wall-bound maltase (α-glucosidase). In baker's yeast the estimation of mural invertase was achieved by preventing hexose metabolism in the presence of fluoride (MEYER and MATILE 1974 b). The application of this technique is restricted by the lack of suitable inhibitors which do not impair the semipermeability of the plasmalemma.

The most direct means of evaluating mural hydrolase activities appears to be the stripping of walls, that is, production of spheroplasts by means of appropriate cell wall lytic enzymes. In *Saccharomyces* the production of spheroplasts through the action of snail gut enzyme is accompanied by the release of invertase, acid phosphatase, glucanases, β-glucosidase (see LAMPEN 1968), naphthyl-esterase (WHEELER and ROSE 1973) and aminopeptidase (MATILE 1969, ROCK and JOHNSON 1970). The mural nature of β-glucosidase in *Mucor rouxii* mentioned above (FLORES-CARREON *et al.* 1970) has also been confirmed by converting germinated spores into spheroplasts.

A morphological mutant of *Neurospora crassa* which lacks cell walls, yet is viable and grows under proper osmotic conditions, releases 95% of its external invertase into the culture medium (BIGGER *et al.* 1972). In contrast, about 25% of the total cellular invertase and trehalase (another mural hydrolase) are present in purified *Neurospora* wall preparations (CHANG and TREVITHICK 1972 b). These facts raise the question of how the mural hydrolases are integrated into the wall.

1.4.2. Binding of Hydrolases to Cell Walls

If activities of patent hydrolases (assayed in intact cells or tissues) are compared with activities measured in isolated walls it may appear that only a fraction of the external enzyme is eventually recovered in the purified wall preparation. An extreme example of this behaviour has been encountered in the petals of *Ipomoea*. Invertase activity assayed in the intact tissue exactly equaled the total activity present in extracts; hence, practically all of the enzyme appeared to be external, nevertheless only traces of invertase were associated with isolated walls (WINKENBACH and MATILE 1970). According to this result the presence of free hydrolase molecules trapped somehow between the plasma membrane and the cell wall is likely. Since the enzyme is not released from the intact cells, the wall appears to represent a molecular sieve with pores narrower than the diameter of the hydrolase. Indeed, some of the cytochemical localizations of acid phosphatase suggest the concentration of this enzyme in a narrow space adjacent to the plasmalemma (*e.g.*, FIGIER 1968, GENEVÈS 1969, CATTESON et CZANINSKI 1968).

Should the molecular sieve theory of hydrolase retention in the cell wall be correct, then the trapping and releasing of macro-molecules would be determined by either the molecular weight of the hydrolases and/or the size of the pores in the wall. TREVITHICK and METZENBERG (1966) investigated this hypothesis extensively in *Neurospora*. This fungus contains two forms of external invertase, a small isoenzyme and a large, aggregated form of this monomer. Both isoenzymes are partially mural and partially released into the medium. In wild type strains a preferential release of the small invertase occurs, whereas morphological, osmotically sensitive mutants possessing altered cell walls with a greater porosity, release comparatively more of the large isoenzyme (Fig. 3). In protoplasts the fractionation of small and large isoenzymes is evidently completely absent. Moreover, the plot in Fig. 3 demonstrates that the fraction of total invertase released into the medium is considerably less in the mould with normal walls as compared with the osmotic mutants.

The proposed model of sieving of molecules by *Neurospora* cell walls seems to explain the fact that hydrolases of low molecular weight such as RNase (m.w. 10,000) and acid proteinase (m.w. 22,000) are not retained in the wall, whereas β-glucosidase (m.w. 168,000), invertase (m.w. of large isoenzyme 210,000) and trehalase (m.w. 300,000) are retained (CHANG and TREVITHICK 1974). The existence of a network of pores, 4–7 nm in diameter, in the *Neurospora* cell wall (MANOCHA and COLVIN 1967) seems to support this view. However, the molecular sieve theory fails to explain the finding that neither by simple washing nor by chemical reagents can the invertase contained in isolated *Neurospora* cell walls be dissociated from them. Invertase is only released upon treatment of the walls with cell wall lytic enzymes (CHANG and TREVITHICK 1970). Hence, the cell wall (in the case of *Neurospora*) represents more than a molecular sieve comparable with the meshwork of a Sephadex bead, rather is it a trap for large enzyme molecules. The hypothesis of CHANG and TREVITHICK (1974) that trapping of external hydrolases is a

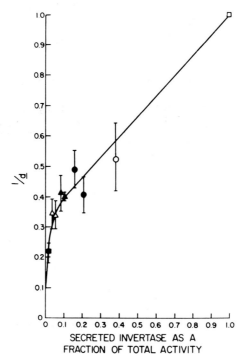

Fig. 3. Molecular sieving of invertase isoenzymes in cell walls of *Neurospora crassa* strains as a function of the fraction of the total invertase secreted. The ordinate, *1/d*, is the reciprocal of the degree of fractionation of small from large invertase; *d* represents the ratio of activities of small invertase to total invertase present in the culture medium. △ wild type; ▲ crisp; ● osmotic mutant of crisp; ○ osmotic mutant; □ spheroplasts of wild type; ■ wild type with cycloheximide. From TREVITHICK and METZENBERG (1966).

consequence of wall rigidification in the growing hyphae (Fig. 4) is obvious, although conclusive evidence has still to be produced.

The cell wall of baker's yeast is characterized by extremely narrow pores; the exclusion volume for polyethyleneglycol corresponds to a molecular weight of only ca. 4,500 (GERHARDT and JUDGE 1964). The external invertase, by contrast, is a large mannan protein with a molecular weight of 270,000. It is not released into the medium nor are other external enzymes released. Despite this evidence for the trapping of invertase by the wall, a covalent binding to cell wall polysaccharides has been proposed by LAMPEN (1968). The decisive experimental evidence for this was that a bacterial phosphomannanase which cleaves mannosidic bonds adjacent to a phosphodiester-linked mannose releases the bulk of mural invertase when allowed to react with intact yeast cells. It was concluded that mannan residues of invertase are covalently linked *via* phosphodiester bridges to the cell wall mannan. However, release of invertase from yeast cells can also be achieved by treatments with reducing agents which are supposed to cleave disulphide bridges in the wall (KIDBY and DAVIES 1970). A model proposed by these authors

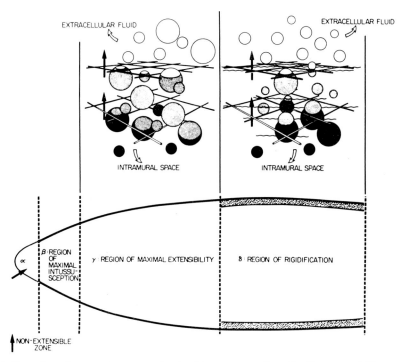

EXTRACELLULAR FLUID   EXTRACELLULAR FLUID

INTRAMURAL SPACE   INTRAMURAL SPACE

β REGION OF MAXIMAL INTUSSU-SCEPTION   γ REGION OF MAXIMAL EXTENSIBILITY   δ REGION OF RIGIDIFICATION

α

NON-EXTENSIBLE ZONE

Fig. 4. Trapping of secreted enzymes in the *Neurospora* cell wall. Courtesy of TREVITHICK.

in which invertase is not chemically bound but physically trapped by the cell wall structure is depicted in Fig. 5. It is evident that stretching of the cell wall by an osmotic "shock" treatment in conjunction with the reducing of the disulphide bridges facilitates the release of mural yeast enzymes (SCHWENCKE *et al.* 1971); in addition to hydrolases previously localized in the mural space of yeast these authors have noted the external location of alkaline pyrophosphatase and Co^{++}-dependent 5'nucleotidase.

Trapping of hydrolases seems to occur also in higher plant cells. HALL and BUTT (1968) recovered 20% of the total acid phosphatase activity in isolated root tip cell walls. However, it may not be feasible to generalize the molecular sieve model of hydrolase retention and release. External hydrolases of higher plant tissues that cannot be detected in isolated cell walls may be trapped between the plasmalemma and the wall or else they may be present in comparatively large pores of the cell wall. If glycerol is used as a non-aqueous medium for the isolation of cell walls, certain hydrolases, which otherwise would be washed out, remain associated with the isolated walls (KIVILAAN *et al.* 1961). SUZUKI and SATO (1973) suggest that ionic bonds may be involved in linking enzyme proteins to charged groups of cell wall polymers. In this case, wall-associated hydrolases can be extracted with saline or buffer (*e.g.*, KEEGSTRA and ALBERSHEIM 1970).

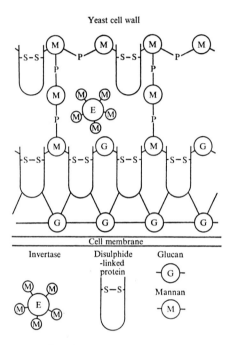

Fig. 5. A model for the yeast cell wall proposed by KIDBY and DAVIES (1970). Invertase is retained within the macromolecular meshwork of mannan-protein. Phosphodiester-linkages between mannose residues and disulphide bridges within the polypeptide moieties of mannan-proteins are important for the trapping of secreted hydrolases.

1.5. Lysosomes

As shown in the preceding sections the space outside the plasmalemma is a prominent constituent of the lytic compartment of plant cells. It is comparatively easy to characterize this space in terms of hydrolases present or absent in cell walls or in the media of submerged cells. In contrast, the localization of hydrolases which appear to be cryptic is difficult. The concept of the lytic compartment requires that unspecific digestive enzymes such as proteinases or nucleases are separated from the sites of the anabolic metabolism by membranes; they are localized in *lysosomes*. The term "lysosome" was coined by DE DUVE (see DE DUVE 1969) as a composition of lysis (dissolution) and soma (body) to designate an organelle present in rat liver cells which contains various digestive enzymes. Its definition was originally based exclusively on biochemical data. In homogenates from rat liver prepared and incubated under isotonic conditions an appreciable proportion of acid phosphatase and other hydrolases is not detectable unless the membranes have been disrupted in one way or another (Fig. 6).

This phenomenon is called *latency*. Latent rat liver hydrolases can be sedimentated; the bulk of them is present in the same fraction as are the mitochondria. It can, however, be separated from these by isopycnic gradient

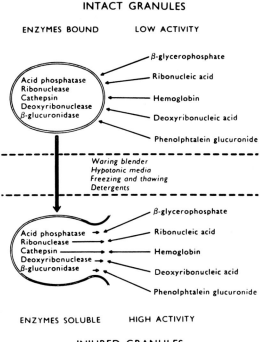

Fig. 6. Biochemical model representing rat-liver lysosomes, as first described in 1955. From DE DUVE (1969).

centrifugation. Hence, DE DUVE's original concept of the lysosome comprehends a number of biochemical properties: a) the presence of several hydrolases in b) a latent and c) sedimentatable form occurring in d) a distinct class of subcellular particles. This concept was later established by electron microscopists who detected the membrane envelope responsible for latency. Moreover, they were able to demonstrate the location of acid phosphatase in rat liver lysosomes cytochemically.

It is now clear that these vacuole-like organelles form a heterogenous class of animal cell structures, cell compartments rather than á species of particles with constant physical and biochemical properties. However, the classical scheme of the rat liver lysosome has misled plant physiologists for a long time. In many cases the employment of the tissue fractionation techniques developed for rat liver lysosomes gave no indication of the existence of similar organelles in plant cells. The reason for these failures is simply that the properties of plant lysosomes may differ greatly from the properties of the rat liver organelles. In plant physiology the cytochemists initiated the discovery of lysosomes.

1.5.1. Cytochemical and Biochemical Approaches

At the level of light microscopy the cytochemical localization of acid phosphatase and many other hydrolases revealed a granular (Plate 4, 5 A), sometimes reticular distribution of the internal activity in various higher plant and fungal cells (e.g., ASHFORD 1970, ASHFORD and MCCULLY 1970, a, b, AVERS and KING 1960, HALL 1969, MCLEAN and GAHAN 1970, MALIK et al. 1969, NEHEMIAH 1973, PITT 1968, PITT and WALKER 1967, REISS 1969, 1971, 1974). Granular distribution in itself would not necessarily imply that the enzyme is compartmentalized; a micrograph presented by REISS (1969) suggests the presence of α-glucosidase in granules of yeast cells (Plate 4 A), yet this enzyme is a constituent of the cytosol (groundplasm). Strong cytochemical evidence for the presence of granular acid phosphatase within a membrane envelope stems from GAHAN's (1965) work showing that granules in root meristem cells are phosphatase inactive unless the tissue has been subjected to treatments that disrupt lipid-protein structures. One of the favourite agents used in this context is the non-ionic detergent Triton x-100 which efficiently destroys cytoplasmic membranes. Hence, the classing of internal, cryptic hydrolases with the lytic compartment or with lysosomes depends on whether or not latency can be demonstrated.

The identification of hydrolase-positive granules with distinct types of lysosomes is difficult if at all possible at the resolution of the light microscope. It will be seen that even the results of electron microscopical hydrolase localizations are sometimes contradictory. In any case, biochemical approaches are necessary to characterize lysosomes. Conclusive data on their enzyme content and other properties can be gathered only if these organelles can be isolated from tissue homogenates.

However, applying the above biochemical criteria for the existence of lysosomes, the examination of homogenates from plant tissues has frequently resulted in an unexpectedly poor sedimentability and absence of latency. In extracts from cotyledons of germinated pea seeds prepared and processed as for rat liver lysosomes only less than 10 per cent of the total acid phosphatase activity was contained in the mitochondrial fraction as compared to some 80 per cent sedimentatable activity in rat liver homogenates; moreover, the small sedimentatable fraction showed no latency (CORBETT and PRICE 1967).

_____

Plate 4. Cytochemical localization of hydrolases in fungi.

A: Saccharomyces cerevisiae: α-glucosidase. ×2,800. B: Aspergillus oryzae: β-glucosidase; note the concentration of the enzyme in the hyphal tip. ×2,800. C: Saccharomyces cerevisiae: leucine aminopeptidase; note the localization in vacuoles (arrows) and near the cell surface. ×2,800. D: Aspergillus nidulans: arylsulfatase; the enzyme is localized in cytoplasmic granules and in the cell walls of hyphae and spores. ×1,400. A–D: Courtesy of J. REISS.

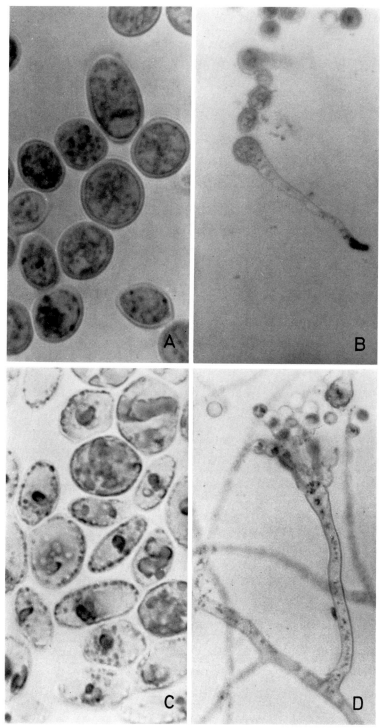

Plate 4.

These and similar results demonstrate the difficulties associated with the isolation of cell structures the properties of which are unknown.

Evidently, sedimentability and latency depend on whether or not the
integrity of possible lysosomes is preserved upon the homogenization of
tissues. They may be disrupted during the disintegration of cells in blendors
or mortars, or else osmotic bursting may occur if the concentration of the
medium in osmotically active solutes is too low. Moreover, the density of the
medium produced by the osmoticum may prevent the lysosomes from
sedimenting (MATILE 1968 a). Other factors may influence the result as well
and last but not least the individual who does the tissue fractionation work
seems to have a considerable effect on the result. This is drastically illustrated
by the results obtained by employing comparable techniques in two laboratories: investigating the sedimenting properties of α-amylase in cell-free
extracts from barley aleurone layers JONES (1972) obtained only traces of
sedimenting enzyme whereas GIBSON and PALEG (1972) succeeded in sedimen-
tating more than one half of the total α-amylase present in the extract. Moreover, the latter authors were able to demonstrate the structural latency of
α-amylase, a prerequisite for postulating its location in a lysosome.

It will be seen that perhaps the crucial point in the isolation of plant lysosomes is the liberation of intact organelles from the cells. Vacuoles represent
a principal class of plant lysosomes and it is obvious that these predominantly
large and fragile structures cannot be obtained by the conventional procedures
of grinding and blending.

1.5.2. Vacuoles

The presence of acid phosphatase in the vacuoles of a variety of plants
including fungi is documented by cytochemical studies (e.g., POUX 1963 a,
1970, CATESSON et CZANINSKI 1968, BERJAK 1968, HALPERIN 1969, VINTÉJOUX
1970, COULOMB 1971 a, HALL and DAVIE 1971, SEXTON et al. 1971, GEZELIUS
1972, FIGIER 1972, CRONSHAW and CHARVAT 1973).

Examples illustrating this type of hydrolase location are given in Plates 3 C,
5 B-D. Considering the extreme unspecificity of acid phosphatase (see
section 1.1.) it is not too surprising that cytochemists are in agreement on its
absence from the cytosol where it would interfere with metabolic processes
such as glycolysis, synthesis of macromolecules etc. In fact, the enzyme seems
to be strictly compartmentalized in living cells. Its release into the cytosol
in late senescence seems to be a specific phenomenon associated with the death
of the cell (BERJAK 1968; see section 3.1.). Practically all the cytochemists
who have observed the vacuolar location of phosphatase have also noted the
presence of the reaction product in other cell structures such as the ER and
dictyosomes. The possible significance of this will be discussed in connection
with the development of vacuoles (section 2.1.). It should be mentioned that
the occurrence of the reaction product of phosphatase in mitochondria,
plastids and nuclei has occasionally been reported. In most of these cases
the necessary control incubations in the absence of substrate or in the presence
of inhibitors such as fluoride were not made. There are also reports in which
the absence of acid phosphatase activity from vacuoles is claimed (e.g., DAU-

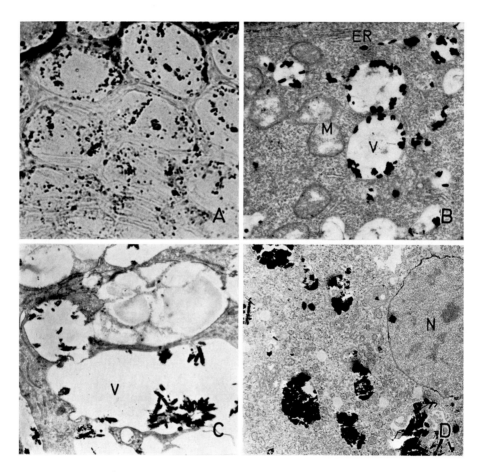

Plate 5. Cytochemical localization of acid phosphatase.

A: Granular localization in cortical cells of barley root meristem. ×515. Courtesy of J. L. HALL. *B:* Product of Gomori-reaction in vacuoles of a root meristem cell of *Cucumis sativus* ×17,500. Courtesy of N. POUX. *C:* Acid phosphatase positive vacuoles in a root cap cell of *Lepidium sativum.* ×13,000. Courtesy of P. BERJAK. *D:* Strong acid phosphatase reaction in vacuoles of an axenically grown myxamoeba of *Perichaena vermicularis.* ×8,500. Courtesy of J. CRONSHAW.

WALDER *et al.* 1969). The causes for such inconsistant results are not clear but the influence of the cytochemical techniques is most likely to be responsible. It appears that in large vacuoles the reaction product is mostly seen adjacent to the tonoplast whereas in small meristematic vacuoles it is spread all over the "cell sap". This phenomenon could be due to attachment of the enzyme protein to a matrix which is necessary for the concentrated deposition of lead phosphate. In any case, further support for the vacuolar location of phosphatase and other hydrolases has been gained from cell fractionation work.

1.5.2.1. Isolation of Vacuoles

As already pointed out, the greatest difficulty in isolating plant vacuoles is the liberation of the intact organelle. The rigidity of the plant cell wall requires drastic procedures to grind up the tissues and, as a consequence, the vacuoles are normally ruptured. This is partially avoided if meristematic tissues, the cells of which contain very small vacuoles, are carefully homogenized (MATILE 1968 a, COULOMB 1968), or if plasmolysed cells of root meristems are disintegrated by chopping (MATILE 1966). Vacuoles with diameters larger than about 2 µm are difficult to obtain by these methods (Plate 6). Upon differential centrifugation these vacuoles sediment together with mitochondria and microbodies. They can be isolated in appropriate density gradient systems according to their densities in the presence of sucrose or Ficoll media (MATILE 1968 a, COULOMB 1968, 1971 b, HEFTMANN 1971, NAKANO and ASAHI 1972, O'DAY 1973). The isolation of root meristem vacuoles by means of discontinuous gradients of sucrose is shown in Fig. 7; it appears that several classes of lysosomes differing in density and in enzyme composition are present in the mitochondrial fraction. The largest vacuoles are trapped on the surface of 15% sucrose.

Rootlets supply a heterogeneous material as far as the size and properties of vacuoles are concerned. The isolation and characterization of vacuoles would be most satisfactory using a tissue with a uniform population of these organelles. Certain laticifers are characterized by the presence of numerous small and uniformly sized vacuoles. Since latex, a liquid cytoplasm, can be obtained by simple tapping of the latex ducts it appears to be an ideal material for isolating vacuoles. In addition, large quantities can be obtained from the rubber tree *Hevea brasiliensis*. It was used by PUJARNISLE (1968) for preparing vacuoles. In spite of the fact that laticifers are highly specialized cells so that information about their lysosomes does not necessarily apply to other plant tissues, the presence of unfused small vacuoles seems to be restricted to the articulated laticifers present in *Papaveraceae*, liguliflorous *Compositae* and certain *Euphorbiaceae*, whereas plants with nonarticulated laticifers have large vacuoles that are destroyed when the laticifers are tapped.

Since the cell wall causes most trouble when liberating intact vacuoles, stripping them and then gently lysing the spheroplasts obtained appears to be a practicable technique. Indeed, efficient procedures for isolating yeast vacuoles from osmotically stabilized spheroplasts have been worked out (MATILE and WIEMKEN 1967, WIEMKEN 1969, WIEMKEN and NURSE 1973).

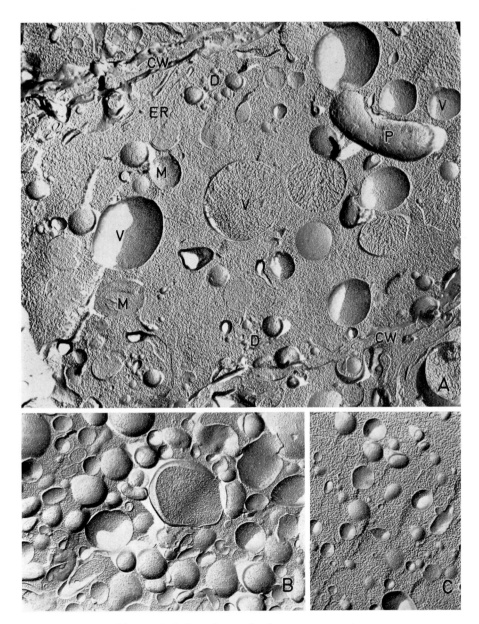

Plate 6. Isolation of vacuoles from corn root tips.

A: General view of a freeze-etched root tip cell containing a number of small vacuoles. ×14,000. *B, C:* Vacuoles obtained from a mitochondrial fraction which was subjected to centrifugation in a discontinuous sucrose gradient (Fig. 6); larger vacuoles were trapped on the surface of 15% sucrose (*B*), smaller vesicles on the surface of 40% sucrose (*C*).
B ×17,800; *C* ×14,100. From MATILE and MOOR (1968) and MATILE (1968 a).

A representation of this technique is given in Fig. 8 and Plate 7 *A–E*; the purity of the preparations of vacuoles obtained appears in Plate 7 *F*. Although the liberation of vacuoles from higher plant spheroplasts has also been reported by COCKING (1960) isolation on a scale which would allow biochemical analysis has not been achieved so far.

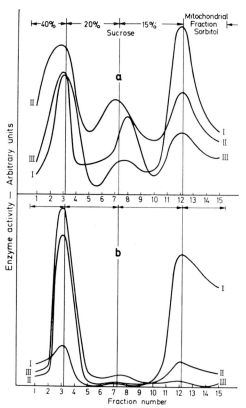

Fig. 7. Subfractionation in a discontinuous sucrose gradient of a mitochondrial preparation obtained by differential centrifugation of a cell free extract from corn root tips. *a* distribution of hydrolases along the gradient. *I*: phosphatase, *II*: proteinase, *III*: RNase. *b* distribution of glutamateoxalacetate transaminase (*I*), NADH₂-dichlorophenol oxido-reductase (*II*) and NADH₂-cytochrome c reductase (*III*). From MATILE (1968).

Yet another technique for separating vacuoles from other cytoplasmic entities is stratification of organelles by centrifugation of whole cells or tissues (JONES 1969, ZALOKAR 1969). Employing this technique of fractionation *in vivo*, vacuoles of subcells of *Acetabularia* have been separated from the cytoplasm and their hydrolase content investigated (LÜSCHER and MATILE 1974).

1.5.2.2. Lysosomal Nature of Isolated Vacuoles

Successful localizations of various hydrolases in vacuoles are summarized in Table 1. Perhaps the most complete analysis of vacuolar hydrolases is available for baker's yeast (Table 2).

Table 1. *Lysosomal Nature of Isolated Vacuoles*

Object	Hydrolases localized	References
Zea mays, root meristem	Proteinase, exopeptidases, RNase, DNase, phosphatase, phosphodiesterase, acetylesterase, β-amylase, α- and β-glucosidase, β-galactosidase	MATILE 1965, 1968 a, unpublished results
Pisum sativum, rootlets	Phosphatase, RNase, α-amylase	NAKANO and ASAKI 1972 HIRAI and ASAKI 1973
Cucurbita pepo, root meristem	Phosphatase	COULOMB 1968
Asplenium fontanum, meristematic fronds	Proteinase, RNase, DNase, phosphatase, phosphodiesterase, β-galactosidase	COULOMB 1971 b
Solanum tuberosum, young dark grown shoots	RNase, phosphatase, phosphodiesterase, acetylesterase	PITT and GALPIN 1973
Lycopersicon esculentum, fruit	Proteinase, RNase, phosphatase	HEFTMANN 1971
Hevea brasiliensis, latex	Proteinase, RNase, DNase, phosphatase, phosphodiesterase, β-glucosidase, β-galactosidase, β-N-acetylglucosaminidase (Peroxidase)	PUJARNISCLE 1968 COUPÉ *et al.* 1972
Chelidonium majus, latex	Proteinase, RNase, phosphatase	MATILE *et al.* 1970
Polysphondylium pallidum	Phosphatase, RNase, DNase, proteinase, amylase, β-N-acetylglucosaminidase	O'DAY 1973
Dictyostelium discoideum, starved myxamoebae	Phosphatase, β-N-acetylglucosaminidase, α-mannosidase	WIENER and ASHWORTH 1970
Acetabularia mediterranea, subcells, prepared from stalks	RNase, phosphatase (Stratification)	LÜSCHER and MATILE 1974
Coprinus lagopus, vegetative hyphae, fruiting bodies	Proteinases, RNase, phosphatase, β-glucosidase, chitinase	ITEN and MATILE 1970
Saccharomyces cerevisiae, spheroplasts	see Table 2	
Neurospora crassa, conidia	Proteinases, aminopeptidase, RNase, phosphodiesterase, phosphatases, invertase	MATILE 1971

A quantitative evaluation of hydrolase distribution in plant cells may be impossible since any procedure that isolates vacuoles and other lysosomal structures is inevitably connected with the loss of an unknown fraction of these structures. In yeast the vacuoles may disrupt during the osmotic lysis of spheroplasts, or else, unlysed spheroplasts may reduce the yield of isolated vacuoles. Therefore, hydrolases (including enzyme present in the external cell space) will always be present in the soluble fraction after isolation of vacuoles. About 50% of the activity of these hydrolases which are possibly

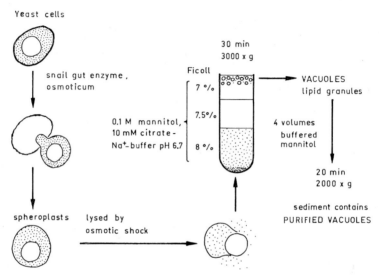

Fig. 8. Schematic representation of the isolation of vacuoles from yeast spheroplasts. Adapted from WIEMKEN and NURSE (1973).

located exclusively in the vacuoles of yeast spheroplasts may be recovered in the fraction of isolated vacuoles, suggesting that in one way or another one half of the vacuoles is lost during cell fractionation. The demonstration of vacuolar locations must, therefore, include the increase in specific activity in the isolate relative to the homogenate and, in addition, the absence of enzyme activities known to be localized in other cell compartments, e.g., in mitochondria or in the cytosol ought to be checked (Table 2).

The structures isolated have not in all cases been morphologically identified as vacuoles. The similarities of techniques employed suggest, however, that vacuole-like organelles have been dealt with. In addition, the latency of hydrolytic activity has not always been checked. As a matter of fact, it cannot be demonstrated if the isolated structure is unstable under the conditions of the enzyme assay. Upon incubation the organelles may gradually become permeable for the substrate (MATILE 1965) or the membranes of yeast vacuoles stretched upon osmotic lysis of spheroplasts may have lost their impermeability for micromolecular substrates (MEYER and MATILE 1974 a). A variety of treatments that result in the disrupture of the lysosomal membranes can

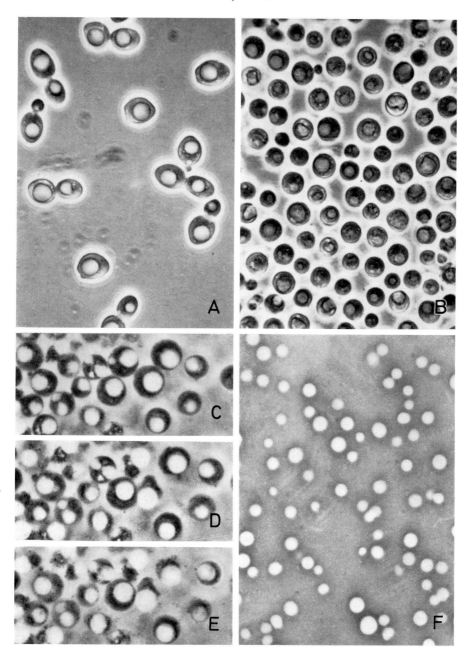

Plate 7. Isolation of vacuoles from baker's yeast cells.

A: Population of budding cells. *B:* Osmotically stabilized spheroplasts. *C–E:* Lysis of spheroplasts in hypotonic medium; the plasmalemma is disrupted whilst the vacuoles remain intact. *F:* Isolated vacuoles. Phase contrast micrographs; ×2,700 (*A, C–E*), ×2,400 (*B*), ×2,000 (*F*). From WIEMKEN and NURSE (1973).

Table 2. *Activities of Enzymes (per Protein) Present in Isolated Yeast Vacuoles as Compared with the Activities in the Total Lysate of Spheroplasts.* The activities in the lysate are taken to be 1 (compiled from MATILE and WIEMKEN 1967, WIEMKEN 1969, VAN DER WILDEN et al. 1973, MEYER and MATILE 1974 a, LENNEY et al. 1974).

Protease A		20.7
Protease B		40.2
Carboxypeptidase		24.0
Aminopeptidase		7.7
Ribonuclease		19.2
Invertase		20.1
α-Mannosidase		20.0
Exo-β-Glucanase		27.8
Alk. Phosphatase (Mg^{2+})	(derepressed cells)	> 40.0
Acid Phosphatase (Co^{2+})	(derepressed cells)	15.0
ATPase (Mg^{2+})	(membrane associated enzyme)	9.0
NADH$_2$-dichlorophenol-indophenol-oxidoreductase	(mitochondrial and microsomal enzyme)	0.638
NADH$_2$-cytochrome c-oxidoreductase	(mitochondrial and microsomal enzyme)	0.016
Cytochrome-c-oxidase	(mitochondrial enzyme)	< 0.01
Succinate-dehydrogenase	(mitochondrial enzyme)	< 0.01
Glucose-6-phosphate-dehydrogenase	(soluble enzyme)	< 0.01
Ethanol-dehydrogenase	(soluble enzyme)	< 0.01
α-Glucosidase	(soluble enzyme, derepressed cells)	< 0.01

be used for the demonstration of latency (Fig. 9). Preferentially, incubations under isotonic conditions at relatively low temperatures for short periods of time are performed for direct demonstration of latency. It appears from Fig. 10 that even under these conditions the absence of substrate accessible to the hydrolase may be demonstrable only for RNase with a high molecular substrate which does not diffuse across the membrane of *Neurospora* vacuoles. On the other hand it is known that the lysosomal membrane of animal cells is permeable for amino acids and even for small peptides; therefore it cannot be taken for granted that vacuolar membranes are always impermeable for small molecules.

Latency of vacuolar hydrolases may be due to other circumstances than the permeability of membranes. Lysosomes of animal cells have been considered as membrane bounded granules with a polyanionic glycolipoprotein matrix to which the cationic hydrolases are bound electrostatically (see KOENIG 1969); according to this model the active sites of hydrolases are inaccessible for interaction with the substrate unless the complex is dissociated.

Hence, the acidic lipoprotein serves as an inhibitor of the lysosomal enzymes, the membrane envelope serving mainly to limit the mobility of hydrolase-lipoprotein complexes within the cell. There is no indication at present for the

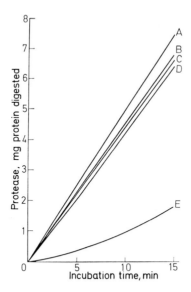

Fig. 9. Latency of sedimentable proteinase present in cell free extracts from tobacco seedlings. *A* incubation in the presence of 0.1% Triton-x-100; *B* ultrasonicated preparation; *C* preparation frozen and thawed repeatedly; *D* incubation in hypotonic medium; *E* untreated control incubated in isotonic medium (20% sucrose). Incubations at 30°. From BALZ (1966).

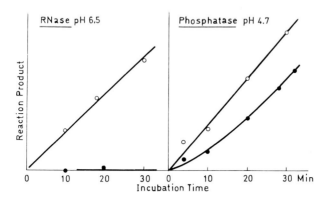

Fig. 10. Latency of acid RNase and phosphatase in vacuoles isolated from germinating macroconidia of *Neurospora crassa*. Untreated (●————●) and ultrasonicated (O————O) vacuoles were incubated at 25° in the presence of substrate containing 0.1 M sorbitol. From MATILE (1971).

existence of such a lipoprotein matrix in plant vacuoles. It is, however, interesting to note that PITT and GALPIN (1973) have detected structural latency of acid phosphatase in potato shoot lysosomes which were treated with detergents and, moreover, they have been able to solubilize this enzyme by treatments at extreme pH values or with phospholipases (Table 3). These findings suggest that phosphatase is in part bound to phospholipids of the lysosomal membrane.

Table 3. *Solubilization of Acid Phosphatase from a Crude Lysosomal Fraction from Potato Shoots Using a Range of Treatments* (data from PITT and GALPIN 1973).

Treatment	% activity of acid phosphatase in fractions		% change in total activity
	Soluble	Particulate	
Freeze/thaw	24.9	75.1	− 24.4
Freeze/thaw + Triton	47.3	52.7	+ 53.0
Ultrasonication	35.6	64.4	− 56.3
Ultrasonication + Triton	63.0	37.0	+ 34.2
pH 2.5	82.9	17.1	− 47.1
pH 10.5	100	0	− 41.1
Venom phospholipases	96.0	4.0	− 38.3
Phospholipase A	24.0	76.0	− 20.4

It appears from Table 2 that enzymes others than hydrolases may be associated with isolated vacuoles. In addition to certain oxidoreductases present in yeast vacuoles (MATILE and WIEMKEN 1967) and meristematic root vacuoles (MATILE 1968 a, HIRAI and ASAHI 1973), vacuolar peroxidase activity has been localized cytochemically (*e.g.,* POUX 1969, CZANINSKI and CATESSON 1970, MARTY 1971 c) and biochemically (COUPÉ *et al.* 1972). Moreover, transaminase activity appears to be present in isolated root meristem vacuoles (Fig. 7; MATILE 1968 a). These non-hydrolytic enzymes are possibly associated with the vacuolar membranes rather than being present in the cell sap. Vacuolar hydrolases may, on the other hand, be localized differentially in the organelle. For instance, RNase is largely solubilized if the isolated yeast vacuoles are disrupted, whereas α-mannosidase is firmly attached to the tonoplast (VAN DER WILDEN *et al.* 1973).

Gel filtration of acid phosphatase solubilized from lysosomal fractions prepared from potato shoots shows that each fraction is characterized by the presence of a distinct pattern of molecular forms of this enzyme (PITT and GALPIN 1973). Fractions of lysosomes from corn root tips which, in density gradients of sucrose, equilibrate at different densities are characterized by distinct patterns of enzyme activities (MATILE 1968 a). Since in the case of higher plants various tissues of an organ contribute to preparations of isolated lysosomes it is difficult to interpret these findings. It is possible that cells or tissues of higher plants contain several distinct classes of lysosomes. However, these classes could also represent developmental stages of vacuoles and it would then have to be considered whether differences in enzyme composition reflect the differentiation of vacuoles in the course of vacuolation.

Vacuoles as a rule contain hydrolases which are known to have also an external localization. In yeast *e.g.,* invertase, β-glucosidase, aminopeptidase have such a dual location. This is a puzzling phenomenon, especially in the case of invertase which is definitely present in yeast vacuoles (BETETA and GASCÓN 1971, MEYER and MATILE 1974 a) yet sucrose, its substrate, does not occur in yeast metabolism. In the case of aminopeptidase, another external enzyme of *Saccharomyces,* the vacuoles contain one distinct molecular form of the four isoenzymes that can be separated by gel-electrophoresis (MATILE *et al.* 1971). An exception is trehalase, an external enzyme of *Neurospora,* which is not located in vacuoles prepared from conidia (MATILE 1971). In contrast to the dual location in yeast of external enzymes, some vacuolar hydrolases such as RNase and endopeptidases may be predominantly or perhaps exclusively internal (MATILE 1969, ROCK and JOHNSON 1970).

It should also be pointed out that not all of the internal hydrolases must be compartmentalized in vacuoles or other lysosomes. An example of this is α-glucosidase which is absent from yeast vacuoles and is most probably located in the cytosol (CORTAT *et al.* 1972). This is also true for the small form of yeast invertase (MEYER and MATILE 1974 a). There seems to be no requirement for compartmentation in the case of these comparatively specific and "harmless" hydrolases. However, reports on the association of RNases with ribosomes or with chromatin do not necessarily reflect the truth about the location *in vivo.* RNases solubilized from vacuoles which were destroyed upon homogenization may bind with their potential substrates, whereby artificial enzyme distributions are produced. In fact, the possibility that RNase is a structural component of plant cytoplasmic ribosomes has recently been ruled out by DYER and PAYNE (1974).

At present the hydrolase inventory of vacuoles is most probably far from being completely known. The most advanced elucidation has perhaps been achieved with regard to proteolytic enzymes associated with yeast vacuoles. It appears from Table 2 that these lysosomes contain a complete set of enzymes for breaking down polypeptides: endopeptidases, aminopeptidase and carboxypeptidase. Indeed, WIEMKEN (1969) has shown that the octapeptide angiotensin II is completely hydrolysed to amino acids in the presence of the vacuolar enzymes. Another interesting fact is that the proteins which specifically inhibit the yeast proteinases are not located in the lysosomes (LENNEY *et al.* 1974). Potentially these proteases seem to be fully active in the living cell.

1.5.3. Aleurone Bodies

An aleurone body is a form of vacuole that is specialized in the storage of seed proteins. They have been studied, therefore, mainly with the aim of identifying specific reserve substances present in these organelles. Using fluorescent antibodies to legumin and vicilin, GRAHAM and GUNNING (1970) were for instance able to localize these major storage proteins of legume seeds in aleurone bodies of bean cotyledon cells (Plate 8). Other reserve proteins have been identified in aleurone grains isolated from a variety of seeds (VARNER and SCHIDLOVSKY 1963, TOMBS 1967, ST. ANGELO et al. 1968, TRONIER et al. 1971). In addition to storage proteins, aleurone grains of certain seed species contain phytate (salt of myo-inositol hexaphosphate). The cytochemical localization of phosphates in reserve cells has shown the concentration of insoluble phosphates in aleurone grains, particularly in the substructure known as a globoid body (POUX 1965). The isolation of globoid bodies from cotton seed aleurone grains has yielded conclusive evidence that phytic acid is one of their principal constituents (LUI and ALTSCHUL 1967). Hence, many of the structural details of aleurone grains can now be interpreted in terms of chemical constituents. It is impossible to cover here the fine structural features of aleurone grains which have been elucidated in the past few years; an example of an aleurone grain with proteinaceous matrix, protein cristalloid, and globoid body is given in Plate 9 A.

From the viewpoint of hydrolase compartmentation aleurone grains deserve special attention, because the metabolically inert storage proteins must be broken down in the course of seed germination. It is evident that this process requires the presence of proteinases within aleurone grains. In fact, corresponding localizations have been successfully carried out in connection with attempts to isolate protein bodies.

There are some difficulties inherent in the isolation of protein bodies. Upon the homogenization of reserve tissues their membranes may be damaged, especially if quiescent seeds are ground up; on the other hand, in germinating seeds aleurone vacuoles may become as fragile as other vacuoles because they gradually lose their compact contents. Hence, water soluble constituents may easily get lost during isolation. Another difficulty in isolating aleurone grains concerns the contaminating particles, especially starch grains which may have similar physical properties and therefore cannot readily be separated by centrifugation. YATSU and JACKS (1968) have avoided these difficulties by selecting an oil storing seed, cotton, and by using a nonaqueous medium, glycerol, for isolating aleurone grains. The result is striking, since all of the acid proteinase and the bulk of acid phosphatase were found to be associated with the isolated aleurone grains. Similar results on the localization of

---

Plate 8. Localization of legumin and vicilin in bean cotyledon cells using fluorescent antibodies.

A–C: Fluorescence photomicrographs of *Vicia faba* cotyledons fixed 63 days after anthesis. *A:* Control without antibodies. *B, C:* Reaction with mixed anti-legumin and anti-vicilin. Arrowed protein bodies are large enough to have been cut open during sectioning; one (*B*) evidently contains antigens, the other (*C*) does not. Both figures include numerous smaller, labelled protein bodies. *D–F:* Sequential staining of the same field of view with: (*D*) anti-legumin; (*E*) anti-vicilin; (*F*) acid fuchsin (bright field microscopy). The enhancement of fluorescence as between (*D*) and (*E*) is shown by most protein bodies (*e.g., l* compared with *l + v*) and indicates the presence of both legumin and vicilin. *v:* aleurone grain containing no detectable legumin. Other protein bodies (such as those encircled in (*F*)) do not react with either antibody. *S* = starch grain. *A, D, E,* and *F:* ×350; *A* and *B:* ×200.
From GRAHAM and GUNNING (1970).

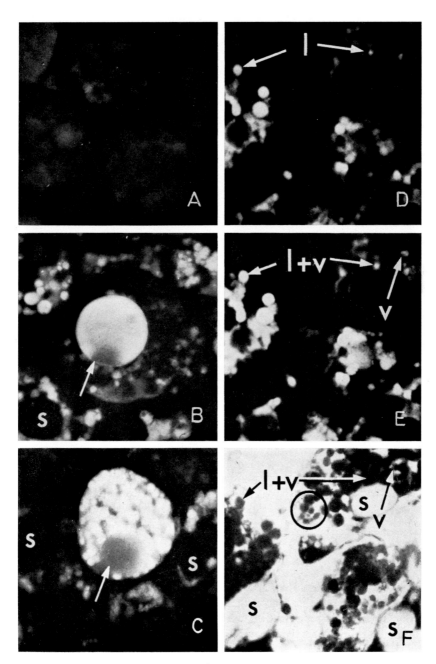

Plate 8.

proteinases, phytase (which is responsible for the mobilization of phytic acid) and other hydrolases in isolated aleurone grains are compiled in Table 4. Cytochemical localizations of acid phosphatase confirm the results obtained with isolated aleurone grains (Plate 9 *B* and *C*).

Table 4. *Lysosomal Nature of Aleurone Grains*

Isolation of aleurone grains

Gossypium hirsutum Cotyledons of ungerminated seeds	acid proteinase, acid phospha- tase	Yatsu and Jacks 1968
Pisum sativum Cotyledons of germinating seeds	Proteinase, RNase, phosphatase, β-amylase, α-glucosidase, acetylesterase, carboxy- and aminopeptidase, phytase	Matile 1968 b, unpublished work
Cannabis sativa dormant seeds	acid proteinase (edestinase)	St. Angelo *et al.* 1969
Hordeum vulgare dormant seeds	specific phytase, acid proteinase	Ory and Henningsen 1969 Tronier *et al.* 1971
Vicia faba germinating seeds	acid phosphatase and -proteinase, phytase	Morris *et al.* 1970
Helianthus annuus cotyledons of germinating seeds	acid proteinase	Schnarrenberger *et al.* 1972

Cytochemical enzyme localizations

Triticum vulgare embryo of germinating seeds	acid phosphatase	Poux 1963 b
Cucumis sativus cotyledons of dormant seeds	acid phosphatase	Poux 1965
Linum usitatissimum	acid phosphatase	Poux 1965
Hordeum vulgare aleurone layer	acid phosphatase	Ashford and Jacobsen 1974

It should be pointed out that the results obtained so far suggest that aleurone grains of dormant seeds contain proteinases together with their potential substrates; this poses the question of how the development of aleurone vacuoles is organized so that these hydrolases do not interfere with the accumulation of storage proteins. Another interesting findings in aleurone vacuoles of pea cotyledons is the presence of hydrolases which apparently have no substrate among the compounds stored in the organelle itself (Matile 1968 b); the presence of enzymes such as RNase and acetylesterase indicates that the aleurone vacuoles are involved in the breakdown of cell components which are localized outside the aleurone grains.

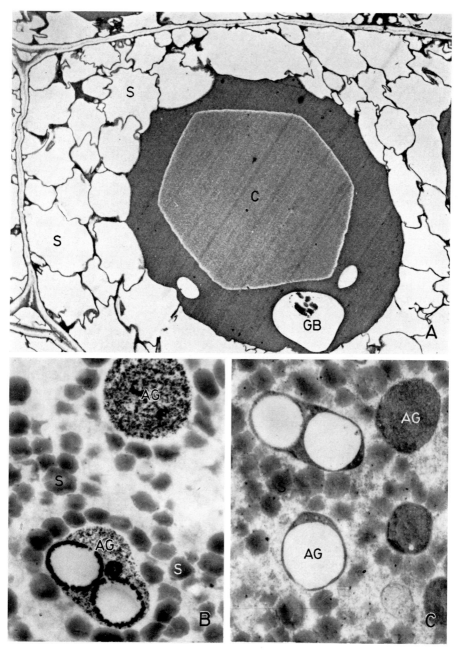

Plate 9.

A: Electron micrograph of a cell from the endosperm of an ungerminated *Ricinus communis* seed. The aleurone grain contains a protein cristalloid (*c*) and globoid bodies (*GB*) embedded in a proteinaceous matrix. Note the numerous spherosomes. $\times$ ca. 4,600. Courtesy of L. Y. YATSU. *B, C:* Localization of acid phosphatase in the embryo of *Triticum vulgare;* reaction product is present in the aleurone grains but absent in the spherosomes (*B*). *C:* Control without substrate. $\times$20,000. Courtesy of N. POUX.

1.5.4. Spherosomes

Although ubiquitous in the plant kingdom and easily recognized in the light microscope by their high refractivity and globular shape, spherosomes are a problematical class of cell organelles. It is generally accepted that they are characterized by a high lipid content (triglycerides) which is responsible for their capacity for accumulating lipophilic stains such as Sudan red or Nile blue. However, staining properties may differ greatly in the small spherosomes of relatively uniform size (diameters < 1 µm) and the much larger oil granules present in reserve tissues of fat storing seeds; SOROKIN (1967) proposed, therefore, to use the term spherosome only for the ubiquitous small lipid granules. However, other properties than staining appear to be different within the group of small spherosomes so that it seems justifiable to use the term *spherosome* for the entire, heterogeneous class of organelles that have in common a spherical shape, a high lipid content, and an anomalous surrounding membrane (see p. 50). Another term, *oleosome*, proposed by YATSU et al. (1971), would also be suitable but has not prevailed.

It is not fully established that spherosomes belong to the lytic compartment. Whether or not hydrolases can be detected in spherosomes seems to depend largely on the method employed. A classical object for the study of spherosomes are epidermal cells of *Allium cepa* bulb scales. Apart from acid phosphatase WAŁEK-CZERNECKA (1962, 1965) localized a variety of other hydrolases in the spherosomes of this tissue by means of cytochemistry. However, YATSU et al. (1971) were not able to detect acid phosphatase activity in spherosomes isolated from onions and other non-oily tissues. It is difficult to interpret such contradictory findings. Very careful comparative staining and enzyme cytochemistry studies in cultured plant cells strongly suggest that acid phosphatase activity is in fact associated with spherosomes (HOLCOMB et al. 1967). Similar results were presented by GORSKA-BRYLASS (1965) on spherosomal hydrolases in pollen grains and pollen tubes. In other cases a confusion with small vacuoles may explain the positive results. The apparent fluctuations in the acid phosphatase activity of spherosomes in guard cells observed by SOROKIN and SOROKIN (1968) in *Campanula persicifolia* could well be due to the fact that vacuoles are very small when the stomata are closed, vacuolar phosphatase being concentrated and cytochemically detectable in small granules that look like spherosomes; in contrast, when guard cells are fully turgescent the vacuoles are expanded and the spherosomes proper appear to be free of acid phosphatase activity. Similarly, spherosomes that react positively in a GOMORI-type of incubation for hydrolases may in fact be identical with small vacuoles (ARMENTROUT et al. 1968, WILSON et al. 1970).

It is interesting to note that in thin sections stained for acid phosphatase, spherosomes always appear to be free of reaction product (Plate 9 B and C). This is particularly evident in tissues containing other structures such as aleurone vacuoles which stain positively (e.g., POUX 1963 b). In contrast, SMITH (1974) demonstrated recently the presence of β-glycerophosphatase in spherosomes of developing cotyledons of *Crambe abyssinica*, large oil bodies present in the same cells appear to be devoid of this hydrolase. In certain fungi spherosomes may be confused with Woronin bodies, as well as with small vacuoles (see WILSON et al. 1970); unless the anomalous spherosomal membrane can be demonstrated in a spherical organelle it must be doubted that fungal spherosomes and lysosomes are identical.

Also work on the hydrolase content of isolated spherosomes has yielded contradictory results. Contamination of spherosomal preparations from homogenates of seedlings with other structures could be responsible for the apparent association of acid phosphatase and other hydrolases with spherosomes (MATILE et al. 1965, BALZ 1966, SEMADENI 1967). However, spherosomes from the oleaginous endosperm of tobacco seeds, that can easily be isolated and purified, contain the bulk of various hydrolases present in this tissue (MATILE and SPICHIGER 1968). Contradictory results may also be due to the heterogeneity of spherosomes with regard to hydrolase content. This seems, indeed, to be so in the case of lipase. The presence of lipase in spherosomes would have a similar significance in the mobilization of spherosomal triglycerides as have proteinases of aleurone vacuoles in the degradation of storage proteins. CHING (1968) reported the association of some of the total lipase activities, both acid and neutral, with spherosomes isolated from germinating seeds of Douglas fir. According to MUTO and BEEVERS (1974) the bulk of acid lipase of castor bean endosperm is present in the fat layer obtained upon the fractionation of homogenates. In contrast, isolated spherosomes from another oleaginous storage tissue, cotyledons of *Arachis hypogaea*, showed practically none of the lipase activity found in cell free extracts (JACKS et al. 1967). An original means of avoiding the technical difficulties in the assessment of lipase activity is the cytochemical demonstration of fatty acids formed when isolated spherosomes of *Ricinus communis* are incubated. In this way some evidence of the presence of lipase in spherosomes was gained (ORY et al. 1968). Indeed, SPICHIGER (1969) showed indirectly that fatty acids are released from endosperm in the course of germination of tobacco seeds.

Recently SCHAFFNER (1974) showed that the spherosomes of baker's yeast differ greatly from spherosomes in higher plants in that they contain phospholipids and sterols but only a small amount of triglycerides.

2. Origin and Development of the Lytic Compartment

The cytoplasmic membranes of plant cells form the boundaries of about a dozen distinct compartments. Except for the inner membranes of mitochondria and plastids, these cytomembranes are ontogenetically related, directly or indirectly, to the endoplasmic reticulum (ER). The flattened cisternae of the dictyosomes originate through the coalescence of ER-derived vesicles and vesicles which are, in turn, pinched off from dictyosomes eventually fuse with the plasmalemma. The Golgi-system and the plasmalemma are, thus, ER-homologous membranes (see MORRÉ et al. 1971). It will be seen that the lytic compartment of plant cells is likewise bounded by ER-homologous membranes: vacuolar membranes seem to be direct products of the ER, and the plasmalemma which is indirectly related to the ER separates the cytoplasm from the external space of the lytic compartment.

Plant cytologists have learned how to interpret the static electron micrographs of developing plant cells in terms of a *flow of membranes* from the ER to the Golgi complex, plasmalemma, vacuoles and other organelles. This flow of membranes is characterized by profound changes of membrane properties which are documented by the fine structural and biochemical differences between homologous membranes. The most conspicuous membrane transformation seems to take place in the Golgi system. As it appears that the ER and the vacuolar membranes are never continuous with the plasmalemma, some sort of *membrane incompatibility* must be achieved during the differentiation of ER membranes in the dictyosomes. The discontinuity of the internal, vacuolar space and the external space, which is the result of membrane transformation in the Golgi apparatus, is important with regard to the functions of the lytic compartments of plant cells.

2.1. Ontogeny and Development of Vacuoles

2.1.1. Vacuolation

Since in 1885 DE VRIES concluded from plasmolysis studies that vacuoles must have a "wall" which he termed tonoplast, the existence of a membrane encircling the vacuolar fluid (cell sap) has only occasionally been doubted. In the era of fine structural plant cytology there was an argument about the origin of this membrane. Vacuoles may be persistent organelles in certain organisms such as yeasts. In contrast to the cyclic development of vacuoles from preexisting vacuoles which occurs in these fungi, in higher plant cells and in cells of filamentous fungi vacuoles develop unidirectionally. In very young meristematic cells of vegetative apices or of proliferating zones of root tips typical profiles of vacuoles cannot be seen in electron micrographs (see *e.g.*, PORTER and MACHADO 1960). Considerable evidence has been accumulated that the vacuoles, which become conspicuous as meristem cells increase in volume and are gradually transformed into parenchyma cells, originate from the ER.

Since cisternae at the distal face of dictyosomes may occasionally be inflated and exhibit vacuole-like profiles, the origin of tonoplasts from Golgi membranes was postulated (MARINOS 1963, UEDA 1966). There are, indeed, examples of vacuoles definitely originating from dictyosomes, namely pulsating vacuoles of certain algae that are involved in the excretion of water

---

Plate 10. Origin and development of vacuoles in root meristems.

A, B: Acid phosphatase localized in provacuoles (*PV*) pinched off from the endoplasmic reticulum; *Lepidium sativum. A* ×19,000; *B* ×39,000. Courtesy of P. BERJAK. *C:* Fusion of a provacuole (*PV*) with a small vacuole. ×46,500. *D:* Fusion of larger vacuoles. Note the protrusion of one vacuole deforming the adjacent vacuole (arrows). ×18,500. *E:* Coalescence of two vacuoles. ×13,600. *F:* Engulfment of a dictyosome-vesicle by the tonoplast. ×46,000. *C–F:* Freeze-etchings of *Zea mays* root tips. From MATILE and MOOR (1968).

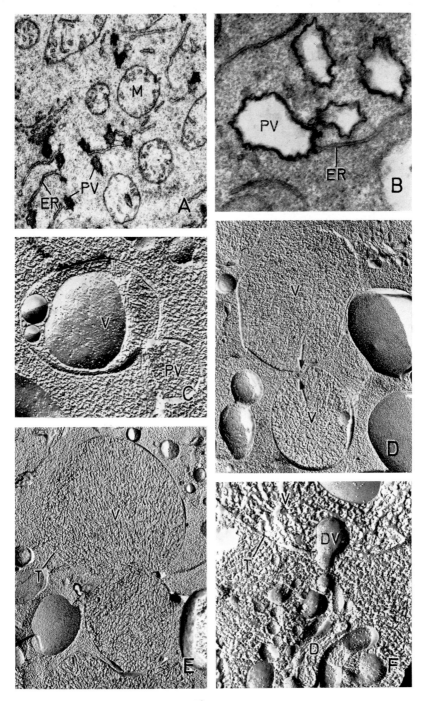

Plate 10.

(*e.g.,* SCHNEPF and KOCH 1966). However, small vacuoles proper originate either by vesiculation of the ER (Plate 10 *A* and *B;* MATILE and MOOR 1968) or else, by the local dilatation of ER-filaments (Plate 11 *A;* BUVAT 1957, BUVAT and MOUSSEAU 1960, POUX 1962 PICKETT-HEAPS 1967 a, MESQUITA 1969, BARCKHAUS 1973). Stellate vacuoles, apparently still connected with the ER, which have often been considered as demonstrating the origin of vacuoles from the reticulum (*e.g.,* BOWES 1965) are most probably artifacts of fixation. FINERAN (1970 a) concluded from comparison of chemical and freeze fixations that "the irregular shapes of vacuoles in thin sections are apparently caused by shrinkage during fixation. When shrinkage is severe, portions of the tonoplast become apposed and superficially resemble profiles of the ER". Small vacuoles pinched off from the ER (subsequently termed *provacuoles*) may exhibit different staining properties with uranylacetate than the ER. This fact gave rise to the opinion that provacuoles are persistent organelles that multiply by division, rather than products of the ER (BARTON 1965). Such differences in electron opacity may, however, merely indicate that membrane differentiation takes place after the fission of ER and pro-vacuoles. The differentiation of ER-membranes at sites of beginning vacuola-tion is in fact suggested by observations of MESQUITA (1969) and POUX (1962): the dilatated regions of the ER are smooth or carry only few ribosomes whilst the juxtaposed regions are rough (Plate 11 *A*).

It appears then that the tonoplast is an individualized and differentiated section of the ER. This view is also supported by circumstancial evidence. For instance, dark staining deposits of tannin, a conspicuous vacuolar com-pound in the shoot apex of *Oenothera* (DIERS *et al.* 1937), and in cultured cells of white spruce (CHAFE and DURZAN 1973) and slash pine (BAUR and WALKINSHAW 1974) appear first in cisternae of the smooth ER which, upon dilatation, gives rise to the formation of tannin vacuoles (Plate 11 *B*). Bio-chemical similarities between ER and vacuolar membranes may strengthen the view that tonoplast and ER are homologous. Fractions of provacuoles and of larger meristematic vacuoles isolated from maize root tips contain enzymes that are regarded as typical membrane constituents of the ER (MATILE 1968 a).

The development of provacuoles into vacuoles comprises the inflation of vacuoles together with extensive fusion of these organelles (*e.g.,* MATILE and MOOR 1968, BERJAK 1972, CHAFE and DURZAN 1973). Profiles of fusing vacuoles can be seen at all stages of vacuolation in maize root meristem; submicroscopic provacuoles, as well as larger vacuoles, show the conspicuous phenomenon of membrane coalescence which seems to be initiated by a protrusion of one vacuole which deforms another (Plate 10 *C–E*). In fully expanded parenchyma cells the numerous globular vacuoles, present in the meristematic condition, have coalesced into a single large vacuole.

Vacuolation as described above is normally an irreversible process. It even-tually ends with the rupture of the tonoplast in the autolyzing cell. But there are some interesting exceptions. In guard cells of opened stomata the vacuoles occupy the majority of the cell space; in *Anemia rotundifolia* HUMBERT and GUYOT (1972) observed the fragmentation and shrinkage of the few large

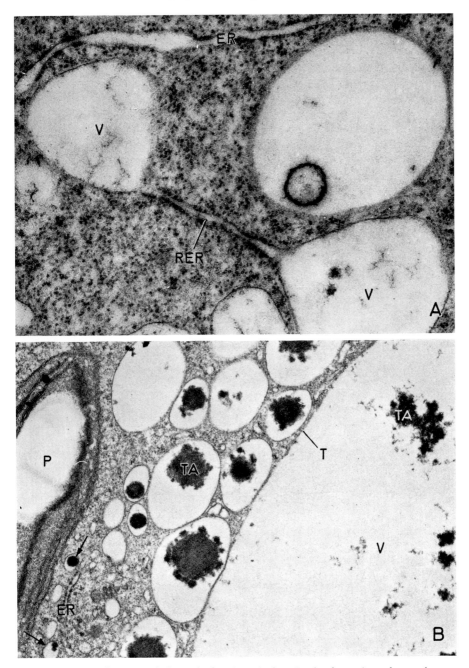

Plate 11. Involvement of the endoplasmic reticulum in the formation of vacuoles.

A: Rough surfaced endoplasmic reticulum linking two developing vacuoles in a root-tip cell of *Lupinus albus.* ×42,000. Courtesy of J. F. MESQUITA. *B:* Portion of cultured white spruce cell showing electron dense tannin materials (*TA*) in vacuoles. Arrows mark tiny tannin containing vacuoles which have arisen from the endoplasmic reticulum. ×26,500. Courtesy of D. J. DURZAN.

vacuoles into numerous small electron dense vacuoles as the stoma aperture is reduced upon the loss of turgor in the guard cells. This process is reversible so that a cyclic change of vacuolar development appears in these highly specialized cells. A similar cyclic development which involves shrinkage and fragmentation on the one hand, inflation and fusion of vacuoles on the other hand, characterizes the budding cycle of yeast (WIEMKEN et al. 1970; Fig. 11). From the viewpoint of the lytic cell compartment a prominent feature of these events is that the tonoplasts are continuous throughout growth, coalescence, and fission of membranes; in other words, compartmentation of hydrolases seems always to be maintained. The rupture of tonoplasts and concomitant

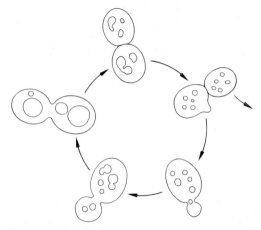

Fig. 11. Cyclic development of vacuoles in the course of the budding cycle of baker's yeast. Adapted from WIEMKEN et al. (1970).

release of hydrolases into the cytoplasm appears to indicate the beginning of autolysis and cell death. In living cells even if highly specialized, as are sieve tubes, the maintenance of compartmentation of hydrolases seems to be strictly observed. This appears from a study of HEYSER (1971) on cell differentiation in the phloem, showing that upon maturation of sieve tubes the central vacuole is deflated and assumes the appearance of a filament of smooth ER. With regard to the metabolic processes residing in the sieve tube "cytoplasm" (as investigated in the phloem exudates) it seems to be important that the tonoplasts, in contrast to HEYSER's (1971) suggestion, are not ruptured in matured sieve tubes. Finally, an interesting case of reversible vacuolation is represented by zygotes of seed plants; as they develop into young embryos the original large vacuoles are reduced in volume and fragmented so that finally the typical vacuole-free structure of meristematic cells is established (SCHULZ and JENSEN 1968, JENSEN 1968).

2.1.2. Involvement of the Golgi-Complex in Vacuolar Development

One prominent role of the Golgi complex in plant cells, its involvement in the production and secretion of cell wall polysaccharides, is documented by abundant cytological, cytochemical and biochemical data (see MOLLEN-

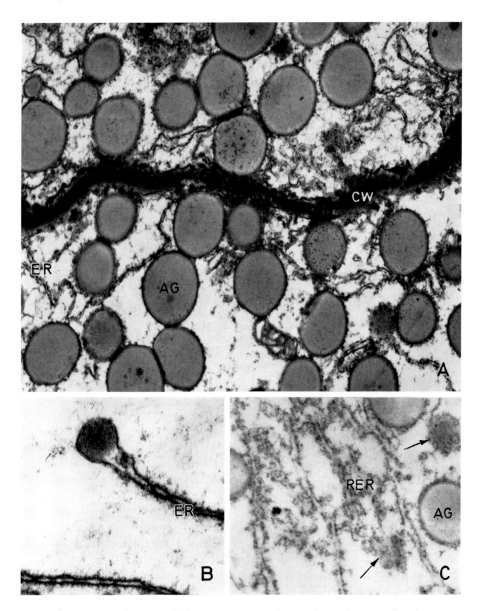

Plate 12. Development of aleurone grains in the starchy endosperm of maize.

A: Electron micrograph of maize endosperm cells 25 days after pollination showing numerous developing aleurone grains. ×18,000. *B:* Protein deposit in an enlarged end of ER. ×57,000. *C:* Ribosomes and polyribosomes on ER and developing aleurone grains; arrows point to near surface sections of granules where polyribosomes are concentrated. ×25,000. From KHOO and WOLF 1970. Courtesy of the Northern Regional Research Laboratory, Agricultural Research Service, U.S. Department of Agriculture.

HAUER and MORRÉ 1966). Vesicles produced by dictyosomes contribute in turn to the development of vacuoles but the significance of this is not yet fully understood.

The incorporation of Golgi vesicles into developing vacuoles has been observed both in thin sections and freeze-etched root tip cells (MATILE and MOOR 1968, BERJAK and VILLIERS 1970, FINERAN 1971, COULOMB and COULON 1971 a) as well as in other objects (BRANDES and BERTINI 1964, PICKETT-HEAPS 1967 b). COULOMB and COULON (1971 a) suggested that this is accomplished by membrane fusion between dictyosome vesicles and tonoplasts. However, it seems rather that these membranes are "incompatible" and that the incorporation of Golgi vesicles is accomplished be means of invaginations of the tonoplast (Plate 10 F); such an "autophagic" uptake into vacuoles (see p. 78) is suggested by a number of electron microscopical observations (PICKETT-HEAPS 1967; MATILE and MOOR 1968, FINERAN 1971). It is likely that the Golgi vesicles incorporated into vacuoles are biochemically distinct from those vesicles which are secreted into cell walls. Differences in size and cytochemical properties in fact suggest differential development of vesicles predetermined either for secretion or for incorporation into vacuoles (MATILE and MOOR 1968, COULOMB and COULON 1971 a). The possibility that the vesicles incorporated into vacuoles bring hydrolases or polysaccharides (which later may provide the mucilage to make the "cell sap" matrix-like) with them should be considered.

2.1.3. Origin and Development of Aleurone Vacuoles

At early stages of development aleurone grains may be indistinguishable from vacuoles (e.g., BUTTROSE 1963, ENGLEMAN 1966). In cotyledons of legumes initial cell expansion is associated with the development of large vacuoles which subsequently divide and differentiate into protein bodies (ÖPIK 1968, BRIARTY et al. 1969). The ontogeny of protein bodies in the starchy endosperm of maize is initiated by the production of provacuoles that are pinched off from the rough surfaced ER (Plate 12 A and B; KHOO and WOLF 1970). These and similar observations in various developing reserve tissues clearly show the vacuolar nature of aleurone bodies.

Evidently the differentiation of aleurone vacuoles comprises the accumulation of reserve proteins (review by MÜNTZ and SCHOLZ 1974). it is not surprising that many authors have reported the involvement of polyribosomes, that are either associated with the membranes of aleurone vacuoles or with the ER, in this process. KHOO and WOLF (1970) presented evidence for the association of polyribosomes with the tonoplast of developing aleurone vacuoles (Plate 12 C). A corresponding observation has been made in Bryopsis; the proteinaceous spherules found in this siphonous green alga develop from swollen cisteranae of the rough ER (BURR and WEST 1971). Synthesis in polyribosomes and subsequent direct transfer of reserve proteins into the cisternae appears to be the logical interpretation of this finding. The synthesis of reserve proteins in other objects appears to be associated with masses of rough ER which can be observed in the proximity of developing aleurone grains (Plate 13; e.g., ÖPIK 1968, BRIARTY et al. 1969). An electron autoradio-

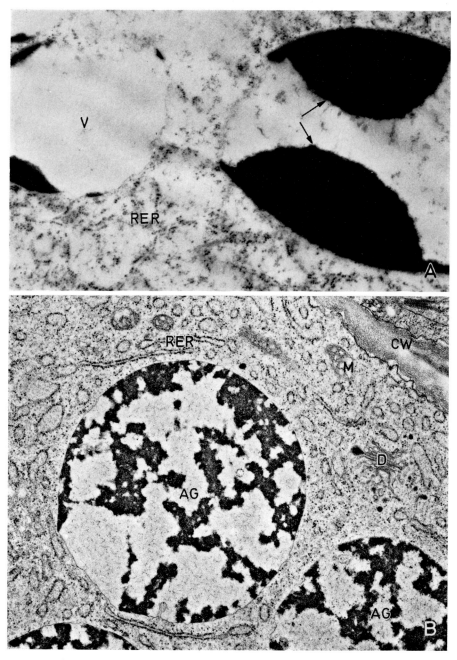

Plate 13. Development of aleurone vacuoles in cotyledons of legumes.

A: Part of a 45 day cotyledon cell of *Vicia faba* with electron dense protein masses in the vacuoles. Note the rough surfaced ER in the vicinity of the aleurone vacuoles. ×13,000. Courtesy of L. G. BRIARTY. *B:* As the pea seeds approach maturity the aleurone vacuoles, which are formed upon the segmentation of the original central vacuole, become spherical in shape and begin to look like the aleurone grains of the dormant seed. × ca. 10,000. Courtesy of H. H. MOLLENHAUER.

graphic study on the synthesis of storage protein in cotyledons of *Vicia faba* has shown that the newly synthesized protein is associated with the rough ER and then moves to the protein bodies (BAILEY *et al.* 1970). Upon chasing the radioactive label in the protein, its transitory localization in the cytoplasm has been observed. This poses the question of how the storage proteins are transported into the vacuolar compartment. Release from the ER and subsequent transport through the tonoplast does not seem to be logical. Although compartmentation is not essential in the case of metabolically inert reserve proteins the involvement of rough ER might suggest that the newly formed proteins are released into the cisterna rather than into the cytosol and transferred to the aleurone vacuoles either through (transitory) connections between ER and vacuoles or by means of ER vesiculation followed by the fusion of the vesicles with the tonoplast. A corresponding phenomenon has in fact been observed in the isolated barley aleurone layer that was induced to undergo changes typical for germination; in the early stages of imbibition the ER vesiculates and vesicles fuse with the tonoplast of aleurone grains (JONES 1969). However, in this case the relationship between the reticulum and the aleurone vacuoles points rather to the synthesis and discharge of hydrolases involved in reserve mobilization than to that of storage proteins.

Aleurone grains isolated from dry seeds contain proteinase. It is difficult to understand the accumulation of reserve protein in the aleurone vacuoles with the degrading enzymes present in the same compartment. In this context it is interesting to note that storage organs are known to contain proteinase inhibitor proteins; some species of seeds contain large amounts of these proteins, some of which are active towards endogenous proteinases whereas others are inactive (review by RYAN 1973). Although it is not known to date whether the inhibitors present in seeds are located in the aleurone vacuoles, it may be speculated that they have a function in these organelles in inhibiting the proteinases during accumulation of reserve proteins. Since they disappear upon seed germination it must be assumed that endogenous proteinases are able to degrade the inhibitors under the conditions in aleurone vacuoles, established upon the inbibition of seeds.

The structure of mature protein bodies has already been mentioned above. Upon germination the proteinaceous matrix and the various inclusions gradu-

---

Plate 14. Origin and development of spherosomes.

A: Micrograph of pea cotyledon late in seed development. Newly formed spherosomes may be closely appressed to the outer surface of the ER (arrows). ×30,000. Courtesy of H. H. MOLLENHAUER. *B:* At high resolution the membrane of a spherosome isolated from tobacco endosperm exhibits a single electron-opaque contour. ×80,000. *C–E:* Development of spherosomes in the *Ricinus* endosperm. Prospherosomes (*PS*) showing the thickening of the middle membrane layer of ER-vesicles (*C* ×81,000; *D* ×92,000). At a later stage of development (*E*) cores bounded by the inner layer of the initial unit membrane can still be seen (arrows). ×20,000, *C–E:* Courtesy of A. SCHWARZENBACH. *F:* Micrograph of a pea cotyledon from a germinating seed. Spherosomes adjacent to the plasmalemma at several stages in the transformation into saccules (arrows). ×100,000. Courtesy of H. H. MOLLENHAUER.

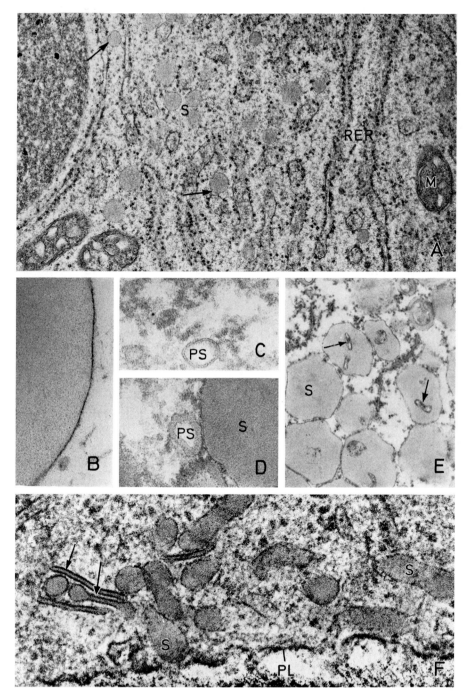

Plate 14.

ally disappear. The aleurone bodies which are comparatively small in the quiescent seed inflate in the course of this process; the aleurone vacuoles coalesce and finally form a single central vacuole (*e.g.*, BAIN and MERCER 1966, TREFFRY *et al.* 1967, VAN DER EB and NIEUWDORP 1967, BRIARTY *et al.* 1970, JONES and PRICE 1970, MOLLENHAUER and TOTTEN 1970, REID and MEIER 1972, SWIFT and O'BRIEN 1972). In reserve organs that die off at the end of seed germination the tonoplast is eventually ruptured and cell autolysis takes place. However, in cotyledons which expand and differentiate into green leaves all the initial aleurone vacuoles eventually form the central vacuole of the mesophyll cells (SIMOLA 1969, GRUBER *et al.* 1970, LOTT and VOLLMER 1973); this again illustrates strikingly the vacuolar nature of aleurone bodies.

2.2. Origin and Development of Spherosomes

The most conspicuous fine structural feature of spherosomes is the membrane. As demonstrated by several authors, these lipid or oil vesicles do not have a normal tripe-layered "unit" membrane. In micrographs of adequate quality only one electron opaque layer can be seen (MOLLENHAUER and TOTTEN 1971, SCHWARZENBACH 1971, YATSU *et al.* 1971, YATSU and JACKS 1972). The anomalous structure of the spherosomal membrane is particularly evident in the isolated organelles (Plate 14 *B*).

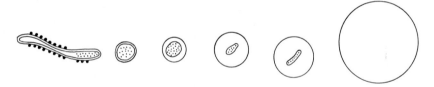

Fig. 12. Origin and development of spherosomes. Adapted from SCHWARZENBACH (1971).

According to FREY-WYSSLING *et al.* (1963) spherosomes originate from the ER (Plate 14 *A*). The reason for the existence of "half unit" membranes in these organelles was detected by SCHWARZENBACH (1971) who followed the development in *Ricinus* endosperm. He observed that the triglycerides are not deposited in the cisternae of the initial ER vesicles but in the middle layer of the membrane (Fig. 12; Plate 14 *C–E*). Upon the development of spherosomes the membrane layer facing the cytoplasm begins to grow and continues to form a membrane-like envelope for the oil body. Remainders of the inner dark layer of the original ER membrane form a core that can eventually be seen in young spherosomes. The peculiarity of the spherosomal membrane seems to have interesting consequences with regard to the further development of these organelles. It is not feasible that spherosomes fuse with structures that have normal membranes and, hence, the mobilization of lipids accumulated in spherosomes must take place either in the spherosome itself or possibly oil bodies may be transferred into vacuoles (see p. 104).

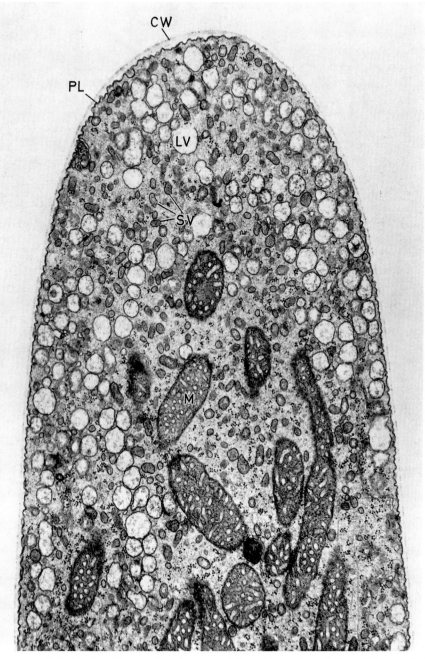

Plate 15. Hyphal tip of *Pythium aphanidermatum*. The apical region of the cell is occupied mainly by vesicles of two general types. The small vesicles (*SV*) have a condensed, densely staining matrix; the larger vesicles (*LV*) have a lighter staining matrix resembling the hyphal wall. ×30,000. Courtesy of S. N. Grove and C. E. Bracker.

2.3. Structures Involved in Hydrolase Secretion

As far as the ultrastructural aspect of secretory processes in plant cells is concerned, almost all the available information is on cell wall synthesis and on the secretion of mucous material. This aspect has been covered by excellent reviews (MOLLENHAUER and MORRÉ 1966, MORRÉ *et al.* 1971) and will only be discussed briefly here. The principal fact is, that the mode of polysaccharide secretion is *granulocrine, i.e.,* it involves vesicles, produced in dictyosomes, the contents of which are discharged into the external space upon the fusion of vesicles with the plasmalemma. Conspicuous secretory activity of dictyosomes is mainly associated with cell plate formation, growing regions of cells (*e.g.,* in tips of root hairs or pollen tubes), or with the production of slime in root caps.

Golgi membranes appear to originate from the ER and the role of the Golgi complex in the membrane differentiation which transforms the original ER membrane into the plasmalemma-compatible membrane of Golgi vesicles has already been discussed. In some fungi, morphologically recognizable dictyosomes, consisting of stacks of flattened and fenestrated cisternae, which are typical for cells of higher plants, may be absent. However, exocytosis of vesicles is a conspicuous phenomenon in the apical tips of filamentous fungi (Plate 15; *e.g.,* GIRBARDT 1969, GROVE *et al.* 1970, GROVE and BRACKER 1970) as well as in bud formation of yeast (MOOR 1967, SENTANDREU and NORTH-COTE 1969, McCULLY and BRACKER 1972). Dictyosomes seem to be replaced in these organisms by transitional membrane elements between the ER and the secretory vesicles.

Cytochemical analysis as well as radioactive labeling experiments have yielded ample evidence for the involvement of the Golgi complex in the synthesis, transport and secretion of polysaccharides. In addition, glucan-synthetase activity has been demonstrated in Golgi membranes isolated from internodes of etiolated pea seedlings (RAY *et al.* 1969). This study provides a basis for the possible involvement of Golgi vesicles not only in the secretion of polysaccharides but also of hydrolases. Unspecific acid phosphatase was found to be associated with isolated Golgi membranes; this result not only confirms previous information stemming from cytochemical work (see DAU-WALDER *et al.* 1969), it could also explain the presence of acid unspecific phosphatase in the mural space of plant cells. However, the functional significance of acid phosphatase is obscure, and it would be much more important to understand the involvement of structures in the secretion of other hydrolases.

The classical example of a secretory tissue that has been thoroughly investigated with regard to biochemical cytology of hydrolase production and release is the exocrine pancreas of mammals. The digestive enzymes that are

---

Plate 16. General view of a barley aleurone cell facing the starchy endosperm after 16 hours of treatment with gibberellic acid. ×6,830. Numerous vesicles of rough surfaced endoplasmic reticulum abound the basal portion of the cell and are seen in detail in the lower inset (×11,000). The apical region contains mainly laminated RER (upper inset; ×12,850). Courtesy of E. VIGIL.

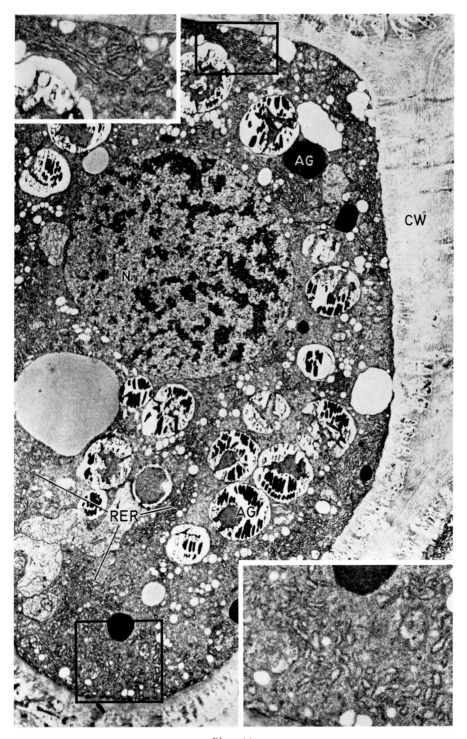

Plate 16.

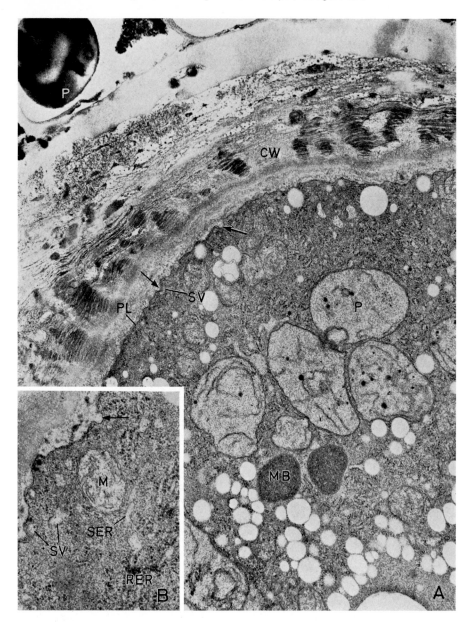

Plate 17. Cross sectional view of a barley aleurone cell adjacent to the starchy endosperm induced to secrete α-amylase (treatment with gibberellic acid for 16 hours).

A: The plasmalemma has a very undulated appearance (arrows); opposite to one of these undulations there is a smooth vesicle (*SV*). ×13,500. *B:* Apparent fusion of smooth vesicles (*SV*) with the plasmalemma (arrows). Note that rough surfaced ER (*RER*) has large regions devoid of ribosomes (*SER*). ×29,400. Courtesy of E. Vigil.

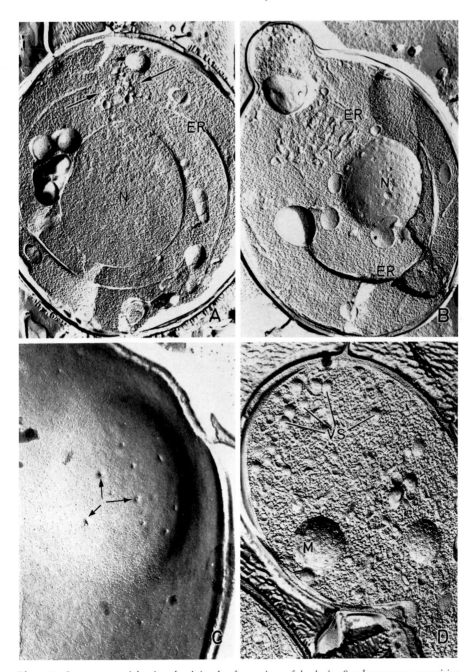

Plate 18. Secretory vesicles involved in the formation of buds in *Saccharomyces cerevisiae*.

A: At the site of the prospective bud numerous vesicles (arrows) are given off by the ER. ×14,000. *B:* Young bud containing numerous vesicles. ×10,000. *C:* Surface view of the plasmalemma of a cell with an emerging bud; arrows point to the sites where vesicles have fused with the plasmalemma. ×30,000. *D:* An older bud containing the ER-derived vesicles (*VS*) and mitochondria. ×26,000. Courtesy of H. Moor.

eventually secreted into the intestine are synthesized in the rough ER (ergasto-plasm); the newly formed proteins are released into the ER cisterna (enchylema) and move into transitional, smooth ER elements via Golgi field and secretory vesicles (zymogen granules) to the external space (see *e.g.*, PALADE *et al.* 1962). An adequate cytological and biochemical analysis of protein secretion in plant cells is not available. However, some of the present rudimentary data suggest the existence of a similar mechanism in the "classical" secretory tissue of higher plants, the graminean aleurone layer. VIGIL and RUDDAT (1972) showed that the induction of hydrolase secretion in isolated barley aleurone layers is associated with the formation of an ergastoplasm-like complex of rough ER sheets. In addition, the vesiculation of smooth ER elements at the secretory pole of aleurone cells appears from micrographs such as presented in Plates 16 and 17. If the secretion of α-amylase in this tissue is inhibited in the presence of actinomycin D, large quantities of this enzyme accumulate intracellularly and the rough ER is transformed into large areas of disarrayed segments of fragmented filaments. These findings clearly demonstrate a correlation between endoplasmic reticulum and secre-tory activity. However, it will only be possible to conclusively demonstrate a pancreatic mechanism of hydrolase release in the aleurone layer after the α-amylase containing membranes which have recently been isolated by GIBSON and PALEG (1972) have been identified. As in barley aleurone cells the activity of secretory cells in insectivorous plants is associated with conspic-uous development of rough ER; SCHNEPF (1963) was, however, not able to observe the formation of vesicles derived from the ER. In *Drosera* and *Pinguicula* the glands secrete slime in addition to proteinases and it would be necessary to demonstrate the presence of these enzymes in prospective secre-tory structures. In another gland, the petiolar nectarium of *Mercurialis annua*, the synthesis of protein in the rough ER, its movement into dictyosomes and secretion into the mural space was investigated radioautographically by FIGIER (1969). That nothing is known about the nature of the exported protein is evidently a drawback. VALDOVINOS *et al.* (1972) were able to correlate the cell separation in the abscission zone of tobacco flower pedicels (which in other objects is known to involve the secretion of cell wall lytic enzymes) with the accumulation of rough ER and the appearence in the abscission cells of Golgi-derived vesicles (Plate 40). In *Achlya ambisexualis* the secretion of cellulase which is associated with the formation of male sex organs (see p. 118) seems to involve Golgi-derived vesicles as shown recently by NOLAN and BAL (1974) employing a cytochemical technique for localizing this enzyme. A detail which could complete the picture of structural involvement in hydro-lase secretion is the budding process in yeast. Vesicles which are present pre-dominantly in budding cells have been isolated. They show high activity of glucanases as well as mannan and mannansynthetase (CORTAT *et al.* 1972, 1973) and seem to be identical with the ER-derived budding vesicles (Plate 18) observed by MOOR (1967).

A vesicular mode of secretion, and the involvement of rough ER and Golgi membranes appears to be logical for three reasons. Firstly, vesicles trans-porting hydrolases to the external space could be responsible for the oriented secretion which has been observed in various instances. Secondly, secreted hydrolases seem to be glycoproteins in nature; at least those which have been purified and analyzed have proved accordingly to contain a certain proportion of carbohydrate. Since the Golgi complex is known as a principal site of polysaccharide synthesis, it is logical to assume that hydrolases synthesized in the rough ER are subsequently glycosilated in the Golgi compartment or in corresponding transition elements. Thirdly, in cases of secretion of unspecific hydrolases such as acid phosphatase or proteinases the compartmentation of these enzymes would seem to be necessary.

Unfortunately, there is practically no firm evidence which supports these hypothetical viewpoints. One of the thoroughly investigated secreted hydrolases, yeast invertase, is a glycoprotein which contains about 50% mannan. LAMPEN (1972) proposed a mechanism of invertase synthesis and release, involving protein synthesis in the ER and glycosilation in (secretory) vesicles. A thorough investigation of the intracellular localization of invertase has, however, yielded not the faintest evidence for the existence of a secretory vesicle containing this enzyme (MEYER and MATILE 1974 a). This is surprising because TKACZ and LAMPEN (1973) reported the local secretion of invertase in the bud. On the other hand invertase is certainly a "harmless" hydrolase; it has no intracellular substrate and need not to be compartmented. The opposite is true in the case of unspecific acid phosphatase; cytochemical observations reported by VAN RIJN (1974) suggest, in fact, the involvement in yeast spheroplasts of vesicles in the secretion of this "harmful" enzyme. In any case, the example of yeast invertase strongly suggests, that the involvement of vesicles in the secretion of hydrolases should not be accepted as a dogma before consistent evidence has been obtained in a given case. The divergent behaviour of three hydrolases, α-amylase, proteinase and RNase, upon differential centrifugation of extracts from barley aleurone layers GIBSON and PALEG 1972) suggests, for instance, that in this tissue α-amylase and proteinase but not RNase may be secreted in a granulocrine manner.

Finally, a report made by myself on the existence of vesicles involved in the secretion of proteinases in *Neurospora crassa* (MATILE 1965) ought to be corrected. A recent reinvestigation has shown, that these vesicles are, in fact, small vacuoles; they contain proteinases with properties typical for the exclusively intracellular enzymes (HEINIGER and MATILE 1974) and structures containing the proteinases released have not yet been discovered.

2.4. Origin and Intracellular Transport of Hydrolases

The morphological features of the development of the lytic compartment suggest that the rough ER is the principal site of synthesis of digestive enzymes. However, definite experimental facts that would support this view are largely lacking.

As indicated in the preceding section, the involvement of membrane systems in hydrolase synthesis and intracellular transport has been largely elucidated in the case of the exocrine pancreas. Likewise, the availability of methods for isolating fractions of rough and smooth ER, Golgi complexes and lysosomes has allowed the investigation of the synthesis, glycosilation and intra-cellular transport of hydrolases in rat kidney (GOLDSTONE and KOENIG 1973). Corresponding biochemical studies in plant cells are more difficult as tissues specialized for lytic processes are rare. Moreover, suitable tissue fractionation techniques are not yet fully developed. The aleurone layer of the graminean endosperm, a tissue specialized in hydrolase secretion, appears to be ideally suited for future investigations of structural involvements in hydrolase synthesis and intracellular transport.

It appears from cell fractionation studies that hydrolases are present in various subcellular structures present in homogenates. This heterogeneous distribution may reflect the flow not only of membranes but also of lytic enzymes in the course of the development of the lytic compartment.

Specifically, the microsomal fractions which contain fragments of ER and Golgi systems are characterized by the presence of hydrolases (MATILE 1968 a, COULOMB 1971 b, NAKANO and ASAHI 1972, PITT and GALPIN 1973). Rough surfaced vesicles isolated from meristematic pea root cells contain acid phos-phatase, RNase and an antimycin A-insensitive NADH-cytochrome-c oxido-reductase which is regarded as a typical enzyme of rough ER membranes (HIRAI and ASAHI 1973). It is logical to assume that these microsomal hydro-lases represent newly synthesized enzyme protein *en route* to the lytic com-partment. Indeed, cytochemical hydrolase localizations demonstrate the existence of several distinct activities in the ER cisternae (*e.g.*, POUX 1966, CATESSON and CZANINSKI 1967, COULOMB 1969, COULOMB *et al.* 1972). How-ever, the presumable movement of these enzymes into the vacuoles could be explained only, if all the cytochemical and biochemical data obtained from different plant tissues were indiscriminately combined. Controversial hydro-lase localizations reported by cytochemists make such a compilatory task very difficult. An autoradiographical study by COULOMB and COULON (1971 b) showing the movement in root meristem cells of newly synthesized protein from ER to dictyosomes, Golgi vesicles, and finally to vacuoles is very enlightening in that it seems to demonstrate the expected analogy of origin and intracellular transport of hydrolases in plant and animal cells. Although this study was carried out under conditions which presumably cause an intensification of lytic processes (anaerobiosis) the results cannot be interpreted in terms of hydrolase synthesis and movement because the nature of the proteins into which radioactive label was incorporated is unknown.

An important point in the elucidation of hydrolase transfer from the ER to vacuoles concerns the role of the Golgy system. COULOMB and coworkers have produced a vast amount of cytochemical data on acid phosphatase localization in *Cucurbita* root meristem which supports the involvement of dictyosomes in the intracellular transport of this enzyme (*e.g.*, COULOMB and COULON 1971 a, COULOMB *et al.* 1972, COULOMB and COULOMB 1973). They come to the conclusion that a distinct type of very small vesicle (diameter ca. 0.1 μm), given off by dictyosomes, is the vehicle of hydrolases *en route* to vacuoles. These vesicles fuse in one way or another to form larger granules (diameter up to 0.5 μm) which have been termed "primary phytolysosomes". It has, however, been observed that primary phytolysosomes in the same tissue originate also from local dilatations of ER cisternae. As biochemical data of the various types of phytolysosomes of *Cucurbita* rootlets are not available it is difficult to evaluate the significance of these findings. Whether or not the intracellular transport of hydrolases is accomplished by dictyosomes could be relevant with regard not only to the glycosilation of these enzymes but also to the nature of the membrane of vesicles predestined to fuse with tonoplasts. Since tonoplasts seem to possess ER-like membranes it is difficult to understand that Golgi-derived vesicles, which are thought to have a plasma-lemma-compatible membrane, should coalesce with vacuolar membranes.

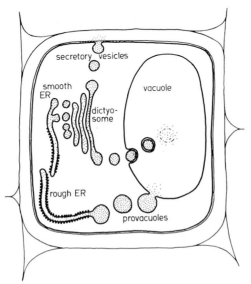

Fig. 13. Tentative diagram of synthesis and intracellular transport of hydrolases in plant cells.

As to the site of hydrolase synthesis the yeast cell is a particularly inter-esting object. Vacuoles originate in a cyclic fashion from preexisting vacuoles (see p. 44). One would expect, therefore, to find a morphological correlate for the necessary supply of hydrolases to vacuoles. However, the only structures which according to MOOR (see MATILE *et al.* 1969) can be observed in intimate contact with yeast vacuoles are the lipid granules which probably contain no hydrolases (SCHAFFNER 1974). Moreover, cell fractionation work has so far yielded no evidence for the existence of hydrolase containing structures others than vacuoles. An obvious explanation of these circumstances would be that in yeasts the vacuolar hydrolases are synthesized in the tonoplast and released directly into the vacuolar fluid. Direct evidence to support this view is not yet available. WIEMKEN (1969) was, however, able to observe polysome-like structures in isolated tonoplasts. As some of the vacuolar hydrolases are glycoproteins the establishment of their synthesis in the vacuolar membrane would indicate that a Golgi step is not essential for glycosilation. Indeed, GOLDSTONE and KOENIG (1973) showed that in rat kidney, glycosilation of hydrolases is initiated in the rough ER. It would be necessary to demonstrate the presence of glycosilating enzymes in the yeast tonoplast. It is not unlikely that the close juxtaposition of enlarging vacuoles and fenestrated sheets of rough ER observed by FINERAN (1973) in freeze-etched living (not chemically prefixed) *Avena* root tips reflects a functional relationship between the two organelles in terms of hydrolase synthesis in the ER and transport into vacuoles. A tentative diagram of synthesis and intracellular transport of hydrolases in plant cells is shown in the Fig. 13.

2.5. The Lytic Compartment of Animal and Plant Cells

In the late nineteen-fifties, cytoplasmic granules which contain several acid hydrolases were isolated from rat liver homogenates by the Louvain group of DE DUVE. DE DUVE (1969) has vividly narrated the history of his important discovery of what he subsequently termed *lysosomes*. Initially the evidence for the existence of lysosomes was based exclusively on biochemical data. The hydrolase-containing granules were characterized by their behaviour upon differential and isopycnic gradient centrifugation and the existence of a membrane wrapping the digestive enzymes was presumed from the latency of these enzymes. Later, when isolated lysosomes were examined in the electron microscope, they appeared as a relatively homogeneous population of membrane-bound granules which were conceived as *the* lysosomes of rat liver cells. Thus, lysosomes seemed to possess physical, biochemical and morphological properties as well defined as those which characterize mitochondria or peroxysomes. In the past decade the original concept of the lysosome as a safety device that allows the storage of digestive enzymes within a living cell has been progressively modified. It has been established that lysosomes have important functions in addition to their role in cell autolysis which occurs upon the liberation of hydrolases from injured lysosomes. On the one hand, lysosomes participate directly or indirectly in a variety of physiological and pathological processes (see comprehensive treatise edited by DINGLE and FELL 1969, DINGLE 1973). On the other hand they are now known as a polymorphous, complex, dynamic, membranous system of animal cells, the classical rat liver lysosome being merely one of its developmental stages. Thus, the lysosome is not a body as its name suggests; it is part of what could be visualized as the lysosomal system. DE DUVE (1969) suggested replacing the term lysosome by the more adequate term *vacuome* to designate the totality of membrane systems involved in cellular lytic processes. Since the term vacuome was introduced by the French plant cytologist P. A. DANGEARD and since moreover, the vacuoles of plant cells undoubtedly are also "lysosomal" in nature, there is no reason to object to its use in animal cytology. On the contrary, it demonstrates the close resemblance of the organization of lytic processes in plant and animal cells.

The diagram presented in the Fig. 14 (JACQUES 1969) shows how in animal cells the vacuome extends from the synthesis of hydrolases and *primary lysosomes* by rough surfaced ER and Golgi complex, through the digestion of endogenous and exogenous material in *secondary lysosomes* (digestive vacuoles), to the formation of residual bodies, the content of which is eventually excreted into the external cell space. Primary lysosomes may either fuse with auto or heterophagosomes, or else, they may function as secretory

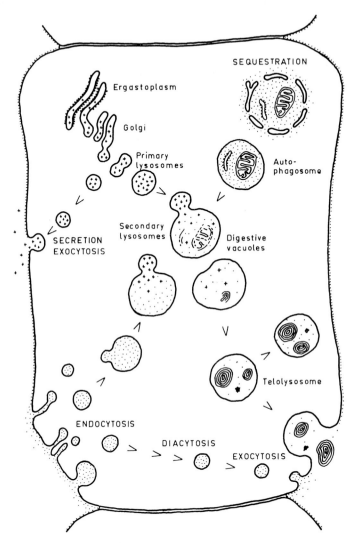

Fig. 14. Schematic representation of the vacuolar system of animal cells. Adapted from JACQUES (1968).

vesicles which fuse with the plasmalemma and release hydrolases into the external space.

DE DUVE (1969) viewed the vacuome "as a system endowed with the same type of continuity as that existing in a chemical factory comprising a complex network of interconnected vats and pipes, controlled by valves that are intermittently opened and shut in a non-synchronous manner". An important feature of the vacuome of animal cells is, that it is continuous (in the above sense) with the extracellular space. A flow of exogenous material within the vacuolar system is realized by *endocytosis* (phagocytosis). A further continuity with the external space is brought about by the discharge of residual bodies. In plant cells, on the basis of present information, endocytosis and formation of heterophagosomes do not occur and this has an important consequence regarding the nature of vacuolar membranes. Undoubtedly, tonoplast and plasmalemma are incompatible membranes. A coalescence of these membranes and, hence, the extracellular discharge of the vacuolar fluid appears to be impossible in plant cells.

The discontinuity between the plant cell vacuome and the external space is necessary because the vacuoles have other important functions in addition to those connected with lytic processes. One of them concerns the storage of large quantities of metabolic intermediates such as organic acids, amino acids and sugars. These are not only responsible for the high turgor pressure which characterizes plant cells; as pointed out by WIEMKEN and NURSE (1973), discussing the significance of storage pools of amino acids present in the yeast vacuole, "the concentrated nutrious cell sap can be regarded as a subcellular analogon of the body fluids in animal organisms". This function of plant cell vacuoles in maintaining homeostasis within the cytoplasm excludes a continuity between the vacuolar fluid and the external cell space. Therefore, plant vacuolar membranes must be incompatible with the plasmalemma. Although detailed biochemical information is still lacking, the membrane differentiation in the Golgi complex which characterizes the ontogeny of animal vacuoles (Fig. 14) apparently does not occur in plant cells.

Exceptional cases of coalescence between tonoplasts and plasmalemma such as observed by GAY et al. (1971) in the development of oospheres in *Saprolegnia* confirm the rule. In this Phycomycete the protoplast of the oogonium is cleaved through the fusion of large vacuoles with the plasmalemma (Plate 19 A) and it must be assumed that the tonoplasts differentiate accordingly before fusion takes place. The cellular slime moulds are another example of lower plant cells which clearly have a vacuolar system continuous with the extracellular space. Not only is the endocytotic uptake of exogenous material a conspicuous feature of the free-living myxamoebae, the contents of autophagic vacuoles may also be ejected into the external cell space (Plate 19 B; GEZELIUS 1972) as is the case in animal cells. In cellular slime moulds the discharge of vacuoles is also a prominent phenomenon associated with spore differentiation (HOHL and HAMAMOTO 1969).

Plate 19. Fusion of vacuoles with the plasmalemma.

A: Developing sporangium of *Saprolegnia ferax*. Transverse section showing vacuoles contiguous in the central region and with extensions which almost contact the plasmalemma (arrow). A nucleus is shown in one lobe of the cytoplasm. ×13,200. Courtesy of J. L. GAY, A. D. GREENWOOD, and I. B. HEATH. *B:* Ejection of the contents of an autophagic vacuole (AV) in the cellular slime mold *Dictyostelium discoideum*. Note the product of Gomori reaction for acid phosphatase present in the vacuoles. ×23,600. Courtesy of K. GEZELIUS.

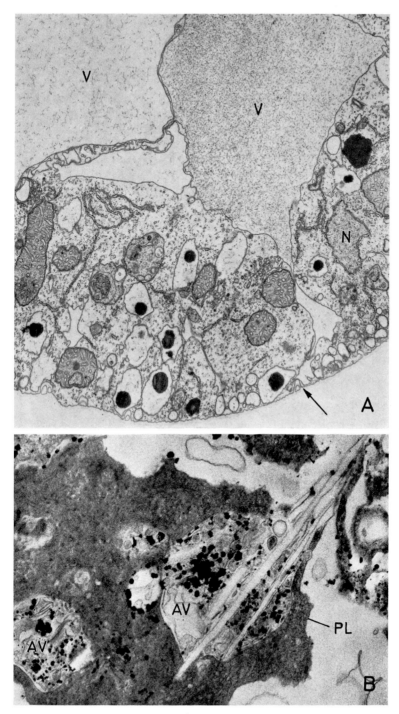

Plate 19.

Apart from these exceptions the plant vacuome is discontinuous with the cell environment. It seems therefore justifiable to use the term vacuome in plant cytology only to designate the totality of vacuoles. DE DUVE (1969) proposed that in animal cytology this term should cover the totality of "lysosomal" cell spaces, intracellular *and* extracellular. In plant cytology it is, in my opinion, preferable to use the term *lytic compartment*. This compartment includes all the cell spaces which are involved in lytic processes. On the one hand the vacuoles and the external cell space are constitutive elements of this compartment as they are the seat of cellular digestive processes; on the other hand the lytic compartment comprise the endoplasmic reticulum and the Golgi complex as being ontogenetically related to the boundaries of vacuoles and external space (tonoplast and plasmalemma) and being probably involved in the synthesis and processing of digestive enzymes. Hence, the main difference between the lytic compartments of plant and animal cells seems to lie in the vacuolar membranes which in animal cells are either homologous with the plasmalemma (heterophagic vacuoles) or, in the case of autophagic vacuoles, must differentiate into a plasmalemma-compatible membrane before exocytosis can take place.

It is uncertain whether in plant cytology a distinction between primary lysosomes, *i.e.,* vtesicles containing digestive enzymes, and secondary lysosomes (digestive vacuoles) is justified. Secretory vesicles which carry hydrolases definitely fall into the category of primary lysosomes. Provacuoles and perhaps Golgi vesicles incorporated into vacuoles can also be viewed as primary lysosomes. It will, however, be seen in section 3.1.3. that vacuolation may occur in such a way that the developmental stage of the primary lysosome is skipped. In yeast, the persistent nature of the vacuoles implies that apart from secretory vesicles only secondary lysosomes are present.

3. Functions of the Lytic Compartment

3.1. Autophagy and Autolysis

3.1.1. Turnover in Biochemical Differentiation

Although it has been recognized that various cellular constituents are metabolically labile, detailed studies on turnover have mainly concerned protein and, to a lesser extent, RNA. This is, of course, due to the importance of protein turnover in biochemical cell differentiation. When a species of protein is synthesized at the expense of amino acids produced by the hydrolysis of other protein species, cells may consequently change their biochemical nature. There are, however, comparatively few examples which demonstrate this particular aspect of protein turnover.

Cotyledons which, upon germination, grow out into green leaves demonstrate most strikingly the role of protein turnover in cellular differentiation. When mustard seedlings are germinated in the light, chloroplasts develop in the mesophyll cells of cotyledons at the expense of storage proteins which are broken down in the aleurone vacuoles. Protein turnover in these organs can, therefore, be described in terms of decline of storage proteins and increase of enzymes and structural proteins of chloroplasts (see HÄCKER 1967).

HALVORSON has shown that specific proteins may be formed in stationary yeast cells in the absence of net protein synthesis (see HALVORSON 1960). In particular, α-glucosidase, induced in the presence of maltose, is synthesized at the expense of the free amino acid pool which in turn is replenished with amino acids from protein degradation. As the inducer is removed from the medium, α-glucosidase gradually disappears which demonstrates that this enzyme is subject to degradation when no longer needed for metabolism. Hence, it appears that cellular proteins may be subjected to continuous degra-

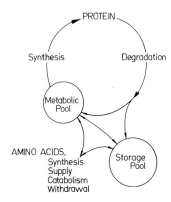

Fig. 15. Tentative diagram of protein turnover and amino acid pools in plant cells.

dation, a constant level merely indicating that breakdown is balanced by continuous synthesis.

It is, however, difficult to analyze protein turnover, that is, to assess rates of protein synthesis and degradation individually (review by HUFFAKER and PETERSON 1974). As indicated in Fig. 15 at least two distinct pools of amino acids are present in plant cells, a storage pool of slowly metabolized amino acids and a metabolic pool which contributes directly to protein synthesis. Attempts to measure net protein synthesis by introducing isotopically labelled amino acids are therefore complicated by unknown interactions between pools of amino acids.

TREWAVAS (1972 a) recently overcame these difficulties by estimating radio-activities of an exogeneously supplied amino acid in a distinct amino acid pool, aminoacyl-transfer RNA, that is, in the immediate protein precursor pool. He was able to calculate from this data the fluxes of labelled methionin in and out of protein either in long term labelling experiments or by chasing the radioactive methionin. Calculations of turnover rates yielded an apparent half-life for protein of about 7 days in fast growing *Lemna*, whereas the lowering of the growth rate by culturing on a suboptimal medium resulted in an increased turnover rate of Protein. An increase in the degradation rate of protein concomitant with a reduced rate of synthesis occurred when growth was stopped by placing *Lemna* on water (TREWAVAS 1972 b).

Hence, it appears that turnover rates reflect the developmental condition of tissues or cells. Low protein turnover rates have, for instance, been determined in growing cultured carrot cells relative to much higher turnover rates in non-growing cells (STEWARD and BIDWELL 1966). In exponentially growing yeast cells protein degradation seems to be immeasurably low, whereas stationary cells are characterized by comparatively high protein turnover rates (HAL-VORSON 1958 a, b). As much as 5% of the total protein, and even more if corrections for experimental errors are made, may be broken down per hour in non-growing yeast cells (FUKUHARA 1967), which would mean that the total protein passes through the free amino acid pool in about a day. Since it must be assumed that only a fraction of all cellular proteins is in fact turned over, the rates of turnover of certain short-lived species of protein must be extremely high. Indeed a study on turnover rates, sophisticated in method, of two metabolically closely related enzymes in cultured tobacco cells, nitrate and nitrite reductase, demonstrates the existence of individual turnover rates; in this study radioactive and density labelling of proteins was combined and it is most interesting to learn that nitrite reductase is comparatively long-lived, its half-life being as long as 124 hours, in contrast to nitrate reductase with a half-life of only 6 hours (ZIELKE and FILNER 1971, KELKER and FILNER 1971). Indirectly, a half-life as short as 2 hours has been determined for a sulfate-permease in *Neurospora crassa* (MARZLUF 1972). This final example demonstrates not only the dynamism of turnover, it also indicates the dynamic nature of membranes the constituents of which, in this case a permease located presumably in the plasmalemma, may be continuously removed and renewed.

Of the various cellular constituents of which the metabolic lability has been assessed or merely demonstrated only the nucleic acids shall be mentioned here in addition to protein. DNA as well as RNA fractions are turned over, both in tissues of higher plants (*e.g.*, HOTTA *et al.* 1965, SAMPSON and DAVIES 1966) and in yeast cells (*e.g.*, HALVORSON 1960, SARKAR and PODDAR 1965).

To make a long and complicated story short, it may be drawn from the various investigations that turnover in plant cells is clearly related to developmental processes. Low turnover rates of protein appear to be characteristic for growth in a culture or tissue that is adapted to largely constant environmental conditions. High turnover rates are, in turn, typical for cells that have become stationary because the environmental conditions have changed *e.g.*, due to the exhaustion of nutrients and the like. Hence, increased rates of protein turnover or degradation indicate that the cells are adapting to the changed conditions. An extreme example is represented by *Euglena* placed on a mineral medium in the dark. Under these conditions this photoauto-trophic flagellate has digested 45% of its cellular protein (55% of the RNA, 30% of the DNA) after 13 days without loosing its viability (BLUM and BUETOW 1963). Starvation appears to induce or increase the degradation of vital macromolecules and this is by far not balanced by the concomitant synthesizing activity. It is interesting to note that this differentiation in starv-ing *Euglena* cells is accompanied by a striking increase of proteinase activity (BERTINI *et al.* 1965). A corresponding observation was made by WIEMKEN (1969) in yeast cells that are either stationary or differentiate in the course of diauxic growth on glucose. Indeed, *Saccharomyces cerevisiae* is an ideal object for studying the relationship between cellular differentiation and hydrolases which are presumably involved in the catabolic phase of turnover reactions. If cultured on glucose this facultative anaerobe among the yeasts

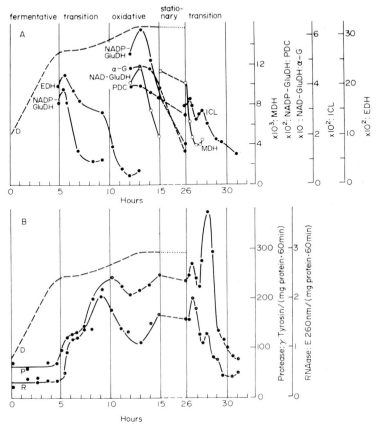

Fig. 16. Diauxic growth of *Saccharomyces cerevisiae* on glucose (*D:* dry cell matter). After 26 hours of cultivation the stationary phase cells were exposed to fresh glucose medium. *A* changes of total enzyme activities during transition phases and stationary growth phase. *EDH* ethanol dehydrogenase; *NAD-* and *NADP-GluDH* glutamate dehydrogenases; *PDC* pyruvate decarboxylase; *α-G* α-glucosidase; *ICL* isocitrate lyase; *MDH* malate dehydrogenase. *B* specific activities of hydrolases; *P* acid proteinase; *R* ribonuclease. From WIEMKEN (1969).

initially ferments sugar to ethanol, oxidative metabolism being repressed in this growth phase. As the glucose is exhausted catabolite repression is relieved and the cells adapt to oxidative metabolism of ethanol. This adaptation takes place in a short stationary phase between the fermentative growth and the subsequent oxidative growth. This so called diauxic lag phase is characterized by the changes of the metabolic machinery, new enzymes are synthesized, others are eliminated. As shown in Fig. 16 *A* similar biochemical alterations take place when the stationary growth phase is reached upon the exhaustion

of ethanol or another nutrient (BECK and v. MEYENBURG 1968). In Fig. 16 *B*
the changes in specific activities of digestive enzymes which occur in the course
of a dauxic growth cycle are shown. The correlations between biochemical
cell differentiation and the activities of proteinases and other hydrolases are
evident and have been thoroughly discussed by WIEMKEN (1969).

The sporulation of yeast cells represents another instance of conspicuous
biochemical differentiation which, according to CHEN and MILLER (1968), is
associated with the enhancement of proteolytic activity.

Measurements of protein turnover give information about the fate of the
average of protein molecules. No distinction between short-lived or relatively
stable species of cellular protein is made, if the corresponding study concerns
merely protein and not a specific species of protein. It would, in fact, be
tempting to follow an individual protein molecule from its synthesis, over
its activity at a given site of the cytoplasm, down to its degradation which,
according to the cellular distribution of the relevant hydrolases, must take
place in the lytic compartment. At the present time such an analysis is not
within the range of even the most sophisticated techniques of cell biology.
Nevertheless, it is interesting at least to ask the question of what determines
the life span of proteins. Definitely, the answer cannot be based on observations *in vitro* because, upon the homogenization of cells, the important aspect
of compartmentation of degrading processes is lost. It seems that the proteolytic inactivation of cytoplasmic enzymes such as tryptophan synthetase in
extracts from yeast described by HOLZER *et al.* (1973) or the proteolytic con-
version in yeast of an inactive form of chitin synthetase into the active form
(CABIB and ULANE 1973) are, in fact, caused by an unspecific endopeptidase
which is localized in the yeast vacuole. Hence, *in vitro* studies cannot contribute to an understanding of what happens in the living cell.

The question of what determines the life-span of protein molecules *in vivo*
is completely open. Obviously it is not only a question of whether adequate
activity of enzymes capable of degrading a given protein are present in cells
but rather a question of what determines whether a protein be transported
to the lytic compartment. Since degradation rates are specific and different
for individual species of protein this transport en route to lysis must also be
specific. One possibility is that the removal from the cytoplasm concerns
specifically "denatured" or "damaged" proteins. Should this be so, then rates
of degradation would be determined by rates of denaturation (the expression
"worn out proteins or structures" covering only our ignorance regarding such
events). In continuation of this speculation it could be hypothesized that
denaturation is not necessarily identical with the loss of the tertiary protein
structure; denaturation could also imply that proteins which are not specifically integrated into multienzyme complexes or into membranes are subjected
to catabolism, the cell being thereby continuously cleared of macromolecules
that are not engaged in metabolism. Conversely, environmental changes or
hormonal effects in growth and development could have the consequence that
the activity of a given species of protein is no longer, or only to a lesser
degree required in metabolism and the protein therefore drops out of multienzyme complexes and structures and is subjected to degradation.

3.1.2. Senescence

Senescence ist the final phase in the development of plant organs ultimately leading to cellular breakdown and death. It is obvious that the main axis of plants, annuals and perennials, has a relatively long life as compared to that of the lateral organs. Leaves have a limited power of growth, after having reached full expansion they sooner or later turn yellow, begin to dry out and eventually abscise. Lower leaves and of course cotyledons may senesce while

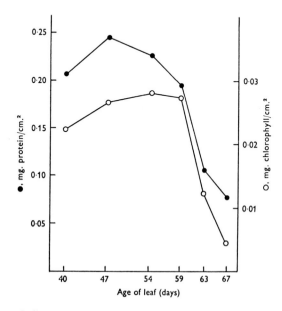

Fig. 17. The chlorophyll and protein content per unit area of the third pair of leaves of *Perilla frutescens,* from the time of completion of expansion to the time of abscission. From WOOLHOUSE (1967).

the plant is still vegetative. Annuals begin to senesce in their entirety when growth is terminated upon flower initiation. Hence, the development of plants is characterized by a conspicuous spatial and temporal pattern of the ageing of their organs.

Life-spans of leaves vary greatly. Conifers have long-lived leaves with life-spans of up to 30 years (*Araucaria*) whereas in annuals, foliar photosynthetic activity may cease after a few months or even weeks. Ephemeral flowers which open and begin to senesce the same day are extreme cases of short-lived plant organs.

At the histological level senescence appears to follow a distinct pattern in the development of tissues. A xylem cell destined to become a tracheid will senesce and differentiate into a dead cell with a specific function long before a parenchyma cell of the same tissue begins to age. Experimental difficulties with metabolic aspects of senescence are due to the differential behaviour of cell types in the tissues. One of the principal features of senescence is the breakdown of cellular material. This is of supreme importance with regard to the economical use of nutrients, especially nitrogen, in plant metabolism. The bulk of total nitrogen is withdrawn from the senescing organs (usually 60–70%) and transported to developing organs or to storage tissues of seeds, tubers and the like.

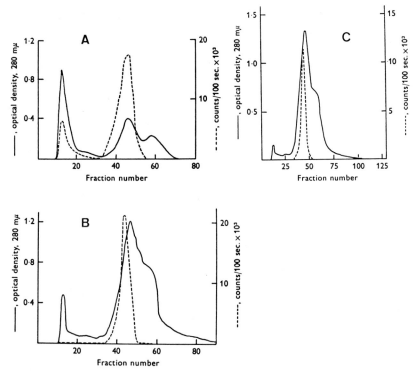

Fig. 18. The soluble protein content and pattern of incorporation of $^{14}CO_2$ into soluble protein of the third pair of leaves of *Perilla frutescens* at different moments of development. Elution diagrams from Sephadex G 100 columns are given for expanding (*A:* 15 days) fully expanded (*B:* 45 days) and senescent leaves (*C:* 65 days). From WOOLHOUSE (1967).

The visible symptom of initiation of senescence, the yellowing of green leaves, is accompanied by the rapid decline of protein content. Of the numerous reports documenting this phenomenon the example of *Perilla frutescens* investigated by WOOLHOUSE (1967) is selected (Fig. 17). The protein content of the third pair of leaves in this short-day labiate which, for experimental reasons was grown under long-day conditions and thus kept in the vegetative state, begins to decrease sharply about three weeks after expansion has been completed. About two-thirds of the maximum protein content, which is reached before the actual onset of yellowing, disappears until the leaves are abscised 67 days after initiation of leaf expansion. Decline of protein content in senescing *Perilla* leaves may, however, not simply be equalled with protein degradation; on the contrary WOOLHOUSE (1967) showed that appreciable protein synthesis takes place until the time of

abscission. Moreover, the pattern of molecular sizes of soluble proteins synthesized in the senescent leaves is different from the pattern in the growing or
mature leaves (Fig. 18). Hence, protein metabolism in a senescent organ may
be viewed as an unbalanced turnover reaction, catabolism exeeding anabolism.
There is fair reason to extend this view to growing organs in which turnover
of protein is unbalanced in favour of synthesis (Fig. 19).

A remarkable feature of the temporal changes of foliar protein in *Perilla* is
that culmination is reached several days before the onset of senescence. A
similar phenomenon was observed by SCHUMACHER (1931/32) in the perianth

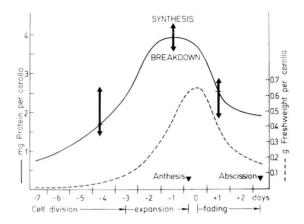

Fig. 19. Flower development of the morning glory *Ipomoea tricolor*. Changes of the total
content of the corolla in protein are viewed as the resultants of changing rates of protein
synthesis and breakdown. Data from WINKENBACH (1969).

of *Phyllocactus* flowers in which the maximum protein content was reached
just before anthesis, considerable breakdown taking place before visible
symptoms of fading appeared. SCHUMACHER, furthermore, noticed an
astonishingly rapid protein breakdown in ephemeral flowers such as those of
Hydrocleis nymphoides in which, upon wilting, more than one third of the
original protein is degraded in one hour. In his day, scientists were not
forced to make their papers as short as possible, so SCHUMACHER, writing in
a dramatic and flowery style, could afford to describe his observations as
follows: „Der Eiweißaufbau ist beim Aufblühen schon beendet, die Maschine
wird umgeschaltet, und während wir uns an der wundervollen Pracht der sich
erschließenden Blüte erfreuen, läuft im Innern das geheime tödliche Spiel der
Eiweißspaltung ab, das, wenn ein gewisser Punkt erreicht ist, nur mit der
Katastrophe des Zusammenbruchs enden kann. Die Geschwindigkeit, mit der
dann eine Spaltung der Eiweiße einsetzt, ist geradezu ungeheuerlich und dürfte
zu den gewaltigsten der bis jetzt im Pflanzenreich bekannten Stoffumsetzun-
gen gehören.“

Indeed, ephemeral flowers offer a unique opportunity to study catabolic
reactions in a senescing organ and, notably the involvement of the lytic cell
compartment. The short-lived corolla of the morning glory *Ipomoea tricolor*

(Plate 20) has been investigated both biochemically and cytologically (MATILE and WINKENBACH 1971). It appears from the temporal changes of protein, RNA, and DNA shown in Fig. 20 A that dramatic catabolic reactions take place in the wilting flower. If these changes are compared with the levels of

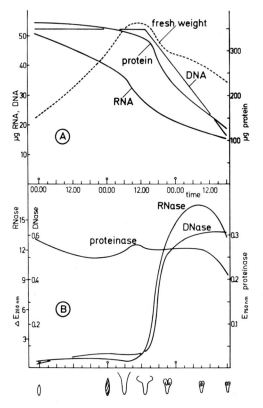

Fig. 20. Senescence in the ephemeral flowers of *Ipomoea tricolor*. Contents in protein and nucleic acids, relative activities of proteinase and nucleases per corolla. The sketches below time axis indicate the shape of corolla. Adapted from MATILE and WINKENBACH (1971).

the corresponding hydrolases (Fig. 20 B) it appears that the breakdown of nucleic acids coincides roughly with a sharp increase in nuclease activity whilst the level of proteinase is practically unchanged throughout senescence. These findings call for two comments. Firstly, the increased activity of certain hydrolases, concomitant with the rapid decline of the overall protein content, exemplifies the above view of unbalanced turnover (Fig. 19) as it has been

---

Plate 20. Development of the ephemeral corolla of *Ipomoea tricolor*.

A: Mature flower bud in the early morning of the day of flowering. *B:* Anthesis at 6 a.m. *C:* Rigid ribs responsible for the shape of the funnel. *D:* First signs of wilting in the early afternoon. *E:* Partially rolled up funnel at 5 p.m. *F:* Corolla in the morning of the day after flowering. From MATILE and WINKENBACH (1971).

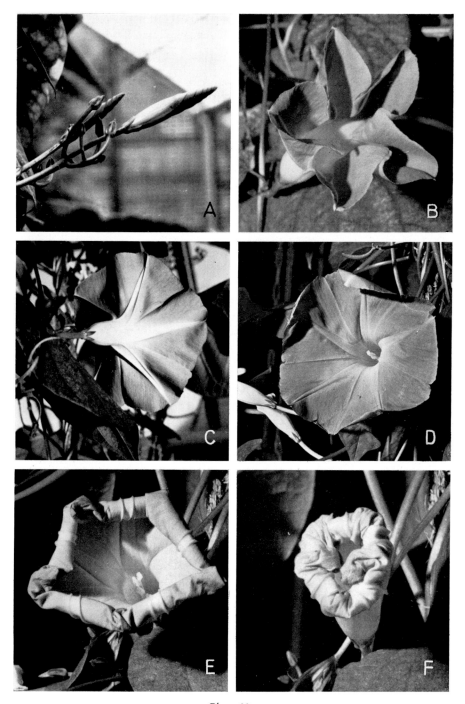

Plate 20.

proved that RNase is synthesized *de novo* in the fading corolla (BAUMGART-
NER *et al.* 1974). The significance of turnover in cellular differentiation is also
illustrated. Secondly, correlating the enzyme activity assayed in crude
extracts with the actual *in vivo* metabolic processes it must be assumed that
proteolysis is increased in the fading corolla. The breakdown of protein is
obviously not controlled by proteinase activity (most probably being compartmentalized) but rather by some process which brings the cellular protein into
contact with the hydrolases. The same conclusion can be drawn from observations on the catabolism of plant cytoplasmic ribosomes reported by PAYNE
and BOULTER (1974).

   Increased RNase activity seems to be typical for senescent leaves (review
by DOVE 1973). This has been observed in the normal ageing of leaves on
the plant as well as in senescence induced by leaf detachment, by environmental stresses, mechanical damage or infection. These factors seem to
accelerate the normal events of senescence in leaves on the plant. Similar
changes in other hydrolytic activities have also been reported: for instance
proteinase increases in detached *Avena* leaves (MARTIN and THIMANN 1972)
and the levels of other hydrolases are raised in ripening fruits (review by
SACHER 1973). The labelling of RNase with deuterium observed by SACHER
and DAVIES (1974) in *Rhoeo* leaf sections that were exposed to D_2O,
unambiguously demonstrates the *de novo* synthesis of this hydrolase in the
senescing tissue.

   Since in senescent tissues metabolism, including protein synthesis, continues
the lytic compartment is probably intact in senescing cells. The breakdown
processes involve, therefore, digestive activity in living cells which is termed
autophagy. As long as the lytic compartment is intact, breakdown is under
control. When it is ruptured the hydrolases mix freely with the remaining
cytoplasm and digestion gets out of control. This final step in senescence is
termed *autolysis* and coincides with cell death.

3.1.3. Autophagy

   Cellular digestion requires the introduction of cytoplasmic material into
the lytic compartment. This implies its transport through lysosomal membranes in such a way that compartmentation of hydrolases is always maintained. In view of this it may be surprising that many electron microscopists
have observed recognizable, though partially digested cell organelles such as
mitochondria in vacuoles. This phenomenon was found by POUX (1936 c) in
young wheat leaves; it was particularly obvious if senescence was induced in
excised fragments of leaves. Abundant evidence for the presence of cytoplasmic material "en voie de dégénerescence dans les vacuoles des cellules
végétales" has been   accumulated since (*e.g.,* SIEVERS 1966, THORNTON
1968, WARDROP 1968, HALPERIN 1969, ZANDONELLA 1970, MATILE and
WINKENBACH 1971, GEZELIUS 1972, CRONSHAW and CHARVAT 1973).
Examples are shown in Plate 21.

   It was noticed already by POUX (1963 c) that inclusions of cytoplasm
wrapped in a membrane are present in vacuoles. This observation provides

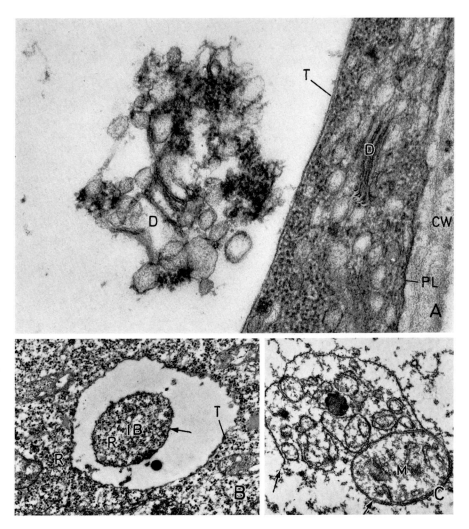

Plate 21. Autophagic activity of vacuoles.

A: Dictyosome (*D*, left side) sequestered in a vacuole of a young *Triticum vulgare* leaf cell. ×65,000. Courtesy of N. Poux. B: A membrane bound intravacuolar body (*IB*) containing ribosomes resembling those in the cytoplasm. ×35,000. *C:* Portion of a large vacuole showing an intravacuolar body containing a mitochondrion and various vesicles; the limiting membrane has ruptured (arrows). ×37,000. *B, C:* Meristematic root cell of *Avena;* Courtesy of B. A. Fineran.

the key to one mechanism of incorporation. In freeze-etched specimens of root tips the membrane which surrounds the vacuolar inclusions has the same fine structural features as the vacuolar membrane (FINERAN 1970 b). In fact, these inclusions are products of tonoplast invaginations which are ultimately pinched off as shown diagrammatically in Fig. 21. It could be claimed that tonoplast invaginations seen in chemically fixed specimens are artifacts,

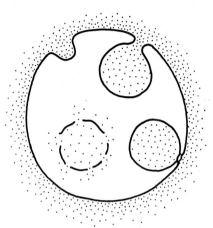

Fig. 21. Autophagic activity of vacuoles: invagination of the tonoplast resulting in the sequestration of cytoplasmic material.

especially when they contain myelin-like whorls of membranes or appear to be empty (THOMAS and ISAAC 1967, COULOMB and BUVAT 1968, BOWES 1969, REID and MEIER 1972, MESQUITA 1972, COULOMB 1973). However, this activity of the tonoplast is equally evident in freeze-etched cells (MATILE 1970, ITEN and MATILE 1970, GRIFFITH 1971). In corn root meristem cells the entire sequence of invagination and formation of intravacuolar vesicles has been observed (Plate 22; MATILE and MOOR 1968). Moreover, FINERAN (1970 b) was able to demonstrate that the fine structure of the freeze-etched tonoplast is remarkably different at sites where the tonoplast invaginates (Plate 22 F); the convex fracture faces of invaginations exhibit a much more concentrated population of globular particles (ca. 10 nm in diameter) as compared to non-invaginating regions of the membrane. Although the function

---

Plate 22. Autophagic activity of vacuoles: invagination of the tonoplast and formation of intravacuolar vesicles in meristematic root cells.

A: Tonoplast with beginning invaginations (arrows) viewed from the inside of a vacuole. Outer fracture face. ×47,000. Courtesy of B. A. FINERAN. B: Beginning invagination (arrow). ×22,500. C: Cross fractured small vacuole with invaginating tonoplast (arrow). ×55,000. D, E: Scars on both intravacuolar vesicle (IV) and tonoplast indicate the sites of membrane fission (arrows). ×27,600 (D); ×25,300 (E). F: Vacuole with beginning invagination viewed from the cytoplasm. At the site of the invagination the inner fracture face is densely populated with globular particles. ×43,600. Courtesy of B. A. FINERAN.

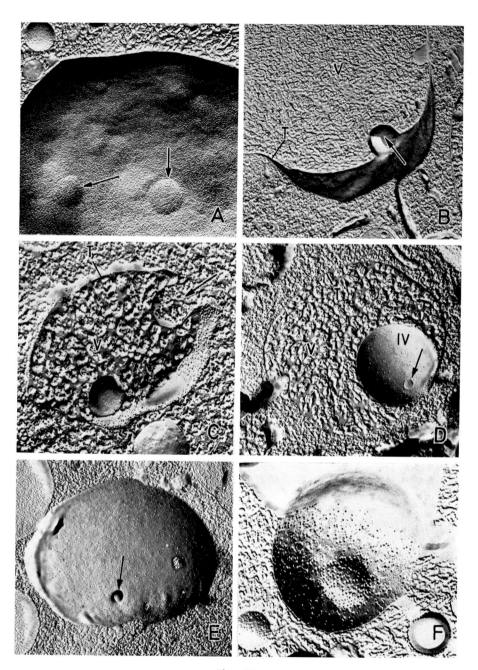

Plate 22.

of these particles is unknown, their crowding at the site of invagination suggests an active role of the tonoplast in the process of engulfment.

Still another remarkable feature of autophagy associated with tonoplast invaginations concerns the nature of the engulfed material. FINERAN (1971) presented electron micrographs showing the specific sequestration of various cell organelles: the vacuolar inclusions in root tips formed by tonoplast invagination contained components of the ER, nuclear envelope, Golgi vesicles, extruded portions of mitochondria and plastids, ribosomes and groundplasm (cytosol). VILLIERS (1971) working with FRAXINUS seeds subjected to prolonged domancy, observed that upon germination the vacuoles in the artificially aged embryonic cells engulfed specifically proplastids and mitochondria. He considers this phenomenon as representing the autophagic elimination of damaged organelles (Plate 23 A, B). A similar behaviour of proplastids was observed in the shoot apices of *Bryophyllum* and *Kalanchöe* (GIFFORD and STEWART 1968). In this case, material is accumulated within the cisternae of proplastid membranes to form large inclusions which are then transferred to vacuoles by tonoplast invagination and protrusion of plastid inclusions. Sequestration of leucoplasts in the central vacuole of *Euphorbia characias* laticifers and convincing evidence that these organelles are wrapped in a membrane which originates from the tonoplast has been reported by MARTY (1971 a). Another example of autophagic uptake of a specific cell constituent, rubber globules, occurs in the vacuolar system of the same object (MARTY 1971 b), though in this case autophagy is perhaps not the right expression since the rubber particles are indigestable and persist in the vacuolar fluid (Plate 23 C).

An event hitherto not satisfactorily explained, associated with the digestion of cytoplasmic material led into the lytic compartment, is the decay of the tonoplast membrane of intravacuolar vesicles which ultimately exposes the sequestered material to attack by the lysosomal hydrolases. This necessary disrupture of vesicles has not only been observed (FINERAN 1971) but can also be derived from the fact that intracellular digestion does indeed occur. However, it is not logical that the tonoplast should be stable as long as it acts as a vacuolar membrane yet becomes labile and disrupt when it surrounds a detached vesicle in the vacuole. One plausible explanation is that the integrity of the tonoplast requires a continuous exchange of membrane con-

---

Plate 23.

A, B: Autophagic activity of vacuoles in the embryo of a *Fraxinus excelsior* seed recovering from pathological changes accumulated over a long period of enforced dormancy. A: Specific engulfment of a proplastid by an invaginating tonoplast. ×15,800. B: Autophagic vacuole with membraneous remains of plastids. ×15,800. Courtesy of T. A. VILLIERS. C: Autophagic uptake of rubber globules into the vacuoles of a laticifer of *Euphorbia characias*. Note the presence of rubber globules in the cytoplasm (single arrow), in intravacuolar vesicles (double arrow) and in the central vacuole (triple arrow). ×41,250. Courtesy of F. MARTY. D: "Cytolysome"-like compartment sequestering cytoplasmic structures in a root meristem cell of a dormant embryo of *Fraxinus excelsior*. Note the double membrane envelope (arrows). ×9,000. Courtesy of T. A. VILLIERS.

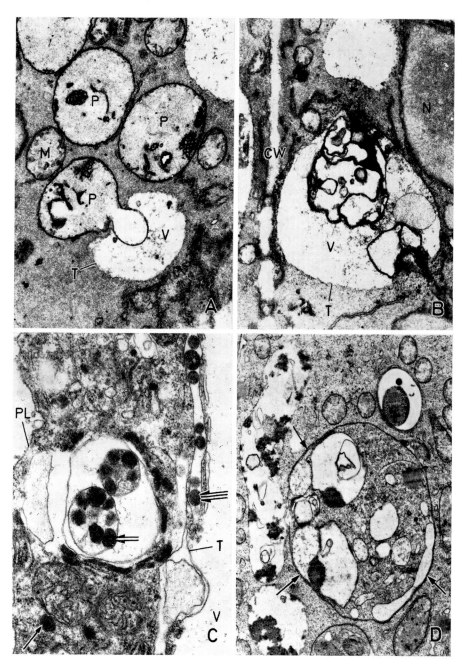

Plate 23.

stituents and perhaps the input of energy furnished by the cytoplasm; such
dependence of the vacuolar membrane on cell metabolism would be inter-
rupted upon the completion of tonoplast invagination and the vacuolar
vesicles would consequently disrupt.

A seemingly different mechanism of autophagy which, anologous to the
phagocytosis-like activity of the tonoplast, provides the continuity of the
lytic compartment throughout sequestration, is documented by the observations of several investigators. VILLIERS (1967) was first to report the presence
of what he termed "cytolysomes", that is, portions of cytoplasm often

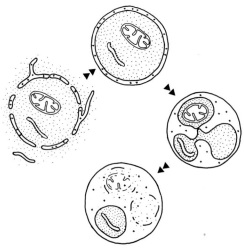

Fig. 22. Formation of autophagic vacuoles: fusion of ER-segments resulting in the
sequestration of cytoplasm.

showing degenerative changes surrounded by two continuous membranes
(Plate 23 D). Acid phosphatase activity present in cytolysomes suggested that
they belong to the lytic cell compartment, and later VILLIERS (1972) was able
to demonstrate that they develop from ER membranes which wrap round
portions of the cytoplasm (Fig. 22). Since VILLIERS' work applied to pathological changes due to artifically prolonged dormancy of *Fraxinus* seeds, it
is important that various other studies *e.g.,* on root meristems and other tissues
show that the sequestration of cytoplasm by the formation of continuous ER
envelopes is a normal autophagic process (*e.g.,* BUVAT 1968, MARTY 1970,
COULOMB and COULOMB 1972, CRESTI *et al.* 1972, MESQUITA 1972). The
ER membranes may occasionally form a series of envelopes so that the cytoplasm is sequestered in a number of concentric shells (Fig. 23; COULOMB and
COULOMB 1972, CRESTI *et al.* 1972, MESQUITA 1972), or numerous ER
vesicles may line up around a cytoplasmic portion, fuse, and finally constitute
a double membrane envelope (BUVAT 1968). In the course of these membrane
building processes, the ER cisterna begins to inflate; after completion of the
membrane envelope the "cytolysomes" closely resemble vacuoles containing
large intravacuolar vesicles (Plate 24). In fact, it is merely the sequence of

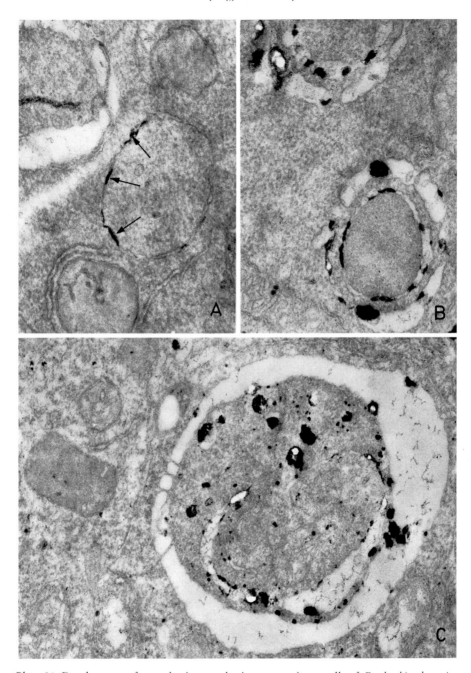

Plate 24. Development of autophagic vacuoles in root meristem cells of *Euphorbia characias*.

A: Formation of membrane envelopes wrapping up portions of cytoplasm. Note reaction product of acid phosphatase in the cisterna of the membrane envelope (arrows). ×45,000. *B:* Concentric membrane envelopes in the course of dilatation. ×45,000. *C:* Dilated envelope; the lysosomal nature of the vacuole formed is demonstrated by the presence of acid phosphatase. ×35,000. Courtesy of F. MARTY.

events that distinguishes the two mechanisms of autophagy: tonoplast invaginations occur *after* vacuoles have developed, whereas in the case of sequestration of cytoplasm by membranes of the ER, vacuolation and autophagy occur *simultaneously*. The close relationship between the two modes of autophagy is furthermore stressed by the fact that both mechanisms may occur in the same tissue. Moreover, CARROLL and CARROLL (1973)

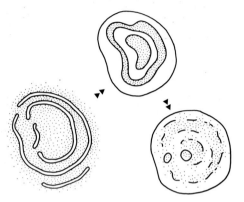

Fig. 23. Formation of autophagic vacuoles: sequestration of cytoplasm involving ER membranes.

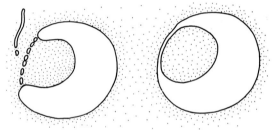

Fig. 24. Autophagy: combination of tonoplast invagination and fusion of ER-derived vesicles in the sequestration of a portion of cytoplasm.

observed a mechanism of autophagic activity in the senescent conidiogenous cell of *Stemphylium botryosum* which represents the exact combination of tonoplast invagination and sequestration by ER membranes (Plate 25; Fig. 24).

In order to demonstrate the sequestration of cytoplasm by ER membranes in detail it would certainly be necessary to produce three-dimensional reconstructions from serial sections. However, the sequence of events can also be constructed from the various stages of formation of autophagic vacuoles which appear in individual thin sections. Moreover, the eventual decay of the sequestered cytoplasm documents that the fusion of ER elements results in the formation of a continuous membrane envelope. Cytochemical studies performed by MARTY (1972, 1973) demonstrate, in addition, the lytic nature of

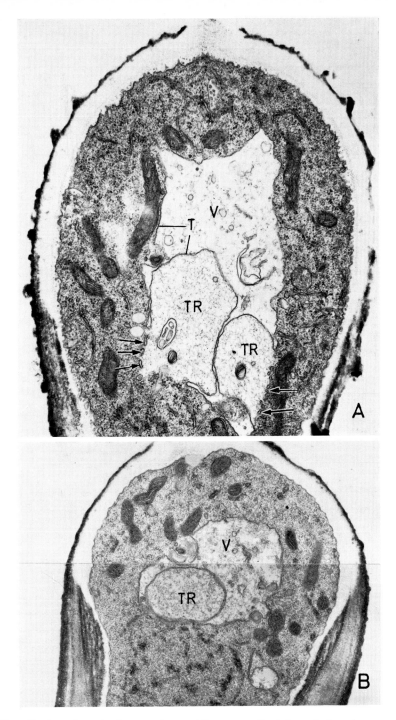

Plate 25. Autophagy in the senescing conidiogenous cell of *Stemphylium botryosum*.

A: Transition regions (*TR*) protruding into the vacuole are being separated from the cytoplasm by a series of vesicles (arrows). ×20,500. *B:* Autophagic vacuole containing membraneous material and a transition region which is possibly separated from the cytoplasm. ×16,800. Courtesy of F. E. and G. C. CARROLL.

these autophagic vacuoles: acid phosphatase appears in the ER vesicles and ER segments during the initial phase of sequestration (Plate 24).

It is important to realize the significance of membrane dynamics associated with cellular lytic processes. In any case, during autophagic activity the strict compartmentation of digestive enzymes appears to be continuously maintained. The portions of cytoplasm that are *en route* to the lysosomal compartment are always first wrapped and sealed into a membrane (tonoplast or endoplasmic reticulum) before they eventually become exposed to the hydrolases. This behaviour emphasizes the importance of the spatial limitation of lytic processes in the living cell and leads to the conclusion that the destruction of the lytic compartment is a feature of cell death.

3.1.4. Autophagy in Cellular Differentiation and Senescence

Autophagic activity is particularly conspicuous in cells which undergo extensive differentiation. A most striking example of enormous autophagic activity associated with cellular differentiation is given by the laticifers of *Euphorbia characias* (MARTY 1970). The majority of the cytoplasm is sequestered through extensive vesiculation of the ER which eventually results in the formation of a large central vacuole containing numerous islands of enclosed cytoplasm (Plate 26). Upon the breakdown of these islands, only a thin layer of cytoplasm between plasma membrane and tonoplast persists. Similarly, meristem cells differentiating into tracheids are cleared of their cytoplasm by autophagy (WODZICKI and BROWN 1973). Another cell development which seems to be associated with extensive vacuolation and autophagy has been observed in the sporangiophore of *Phycomyces* (THORNTON 1968). In cells like these the life-span is short and autophagy may indicate the beginning of senescence. This has, indeed, been shown in the development of the sporogenous cells of *Stemphylium* (CARROLL and CARROLL 1973).

The significance of autophagy in root tips and other meristematic tissues is not clear, as the biochemical correlate of this process has not been elucidated. It can only be speculated that autophagy is the morphological counterpart of what in biochemical investigations appears as the degradative phase of turnover. It is likely that the biochemical machinery of meristematic cells differs greatly from that of the various types of differentiated cells. Hence, the melting down of enzymes and structures in digestive processes seems to have some significance regarding the requirements of differentiation. However, determining the biochemical complement of autophagic activity is a difficult task. In organs of higher plants this is mainly due to the differing behaviour of the cells of various tissues. This appears for instance from electron microscopical studies of the senescing corolla of *Ipomoea* (MATILE and WINKENBACH 1971) and of mesophyll cells of excised portions of *Nicotiana glutionosa* leaves (RAGETLI et al. 1970).

A convincing correlation between autophagic activity and cellular differentiation has been established in *Euglena*. This photoautotrophic organism adapts to autophagic metabolism if it is kept in the dark on an inorganic medium (BLUM and BUETOW 1963, see p. 66). This enabled the demonstra-

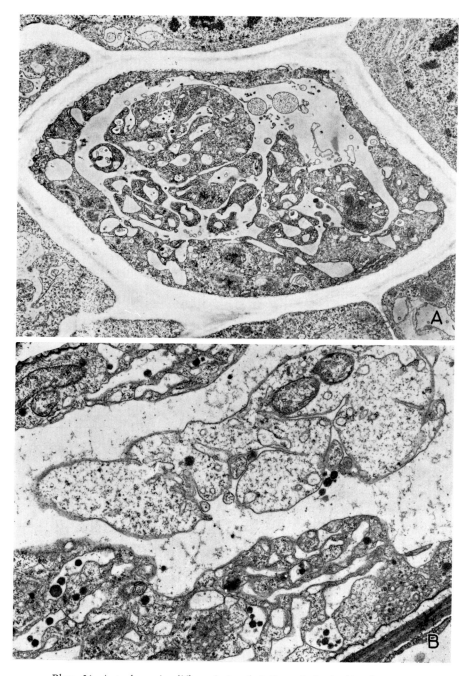

Plate 26. Autophagy in differentiating laticifers of *Euphorbia characias.*

A: Cross section of a laticifer the central cytoplasm of which has been fragmented. Autophagy is associated with the formation of a large central vacuole. ×7,350. *B:* Longitudinal section showing the decay of the cytoplasmic material sequestered in the central vacuole. ×15,000. Courtesy of F. MARTY.

tion of intracellular digestion of cytoplasmic constituents in the electron microscope (MALKOFF and BUETOW 1964).

In yeast cells mitochondria have never been seen in vacuoles, yet changes in respiratory capacity in facultative anaerobes like *Saccharomyces* would offer a unique opportunity to correlate lytic activity with biochemical differentiation. There is small doubt that respiratory enzymes are degraded upon the repression of oxidative metabolism in the presence of high concentrations of glucose. It seems, however, that yeast mitochondria are partially dismantled rather than entirely digested, so that promitochondria persist in the cells in which respiration has been repressed (see MATILE *et al.* 1969). Although invaginations of the tonoplast are noticeable (WIEMKEN 1969) they are not particularly frequent when the respiratory enzymes are repressed in aerobic cells that are transferred to glucose medium. Only such a drastic treatment as the exposure of yeast cells to mutagenic agents, such as acridine orange which causes an extensive breakdown of cellular protein (BOGEN and KESER 1954), induces particularly conspicuous autophagy of tonoplasts (Plate 27; MATILE 1970). Autophagy can also be induced by short exposures to high sublethal temperatures; in this case the tonoplast invaginations almost invariably contain a few lipid bodies (MATILE and MOOR, unpublished results). The significance of this sequestration of yeast spherosomes is unknown.

The existence of specific turnover rates for individual proteins suggests that autophagy is not a random process. It is, however, completely unknown how phagocytosis of proteins or structures to be broken down is triggered. How for instance are the "damaged" proplastids in VILLIERS (1971) studies on aged *Fraxinus* seeds distinguished from the "healthy" organelles? A possible method of recognition of cell components to be eliminated could be some sort of denaturation of protein molecules.

3.1.5. Heterophagy in Plant Cells?

Organisms such as cellular slime moulds, which have typical features of plants as well as of animals, occupy an intermediate position also with regard to the lytic compartment. They resemble the protozoa in that they are capable of digesting food particles intracellularly. When myxamoebae are grown on bacteria, heterophagic uptake of the particulate food and subsequent digestion in food vacuoles take place. Cytological and biochemical aspects of this phenomenon have been studied (*e.g.* HOHL 1965, WIENER and ASHWORTH 1970, GEZELIUS 1971, CRONSHAW and CHARVAT 1973). Exceptional

---

Plate 27. Uptake of cytoplasmic material into vacuoles of acridine orange (*AO*) treated *Saccharomyces* cells. Conditions of treatment: 10 μg/ml AO, 2.4 × 10^8 cells/ml.

A: Beginning invagination of the tonoplast (arrow) in a cell treated with AO for 4 minutes × 17,000. *B:* Tonoplast encapsulating a portion of cytoplasm with a number of vesicles; 10 minutes after addition of AO. ×18,500. *C:* Vacuole containing numerous intravacuolar vesicles in a cell treated with AO for 8 minutes; an invagination extending deeply into the cell sap is visible (arrows). ×33,000. From MATILE (1970).

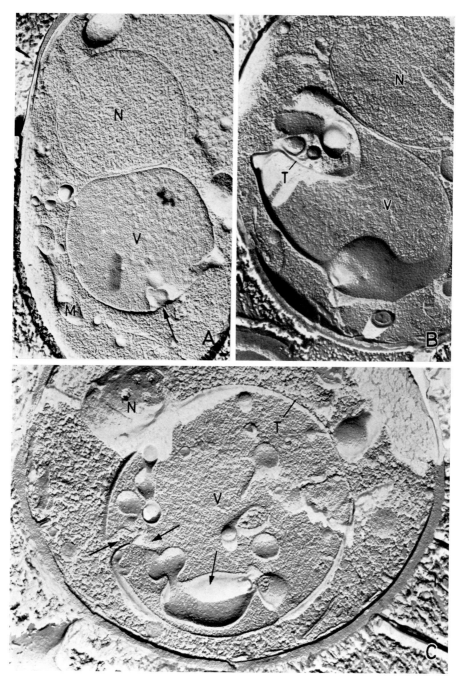

Plate 27.

cases of heterophagy observed in cells of higher plants cannot be directly compared with heterophagy in myxamoebae. An example of apparent heterophagy which involves bacteria is given by the root nodules of legumes. TRUCHET and COULOMB (1973) have shown that the eventual degeneration of *Rhizobium* bacteroids in the senescing part of the nodule is associated with

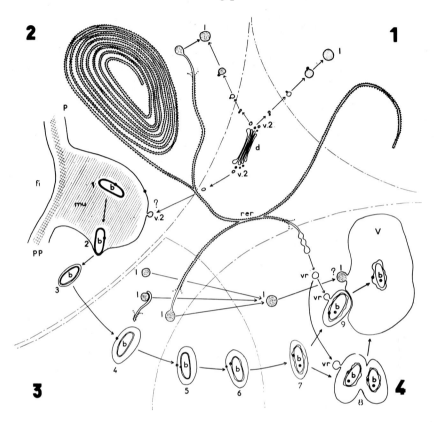

Fig. 25. Formation of heterophagic vacuoles in root nodules of pea. The four sectors of the diagram correspond to the different zones of the nodule. Infection of cells takes place in zone 2. The bacteroids (*b*) develop in the leghemoglobin-containing zone 3. In the necrotic zone (4) the vesicles containing the bacteroids fuse with primary lysosomes (*1, vr*) and are thereby transformed into heterophagic vacuoles. From TRUCHET and COULOMB (1973).

the release of acid phosphatase from primary lysosomes into the membrane-bound compartment in which the symbionts remain throughout the development of the nodules. Since the infection of the host cells involves a phagocytosis-like uptake of the bacteria from the infection threads, the vesicles which contain the bacteroids may be conceived as a kind of heterophagosome (Plate 28). However, the transformation of these structures into digestive vacuoles occurs only after the system has gone through a symbiotic stage of development. The diagram of Fig. 25 summarizes the observations on the involvement of ER and primary lysosomes in this transformation (TRUCHET

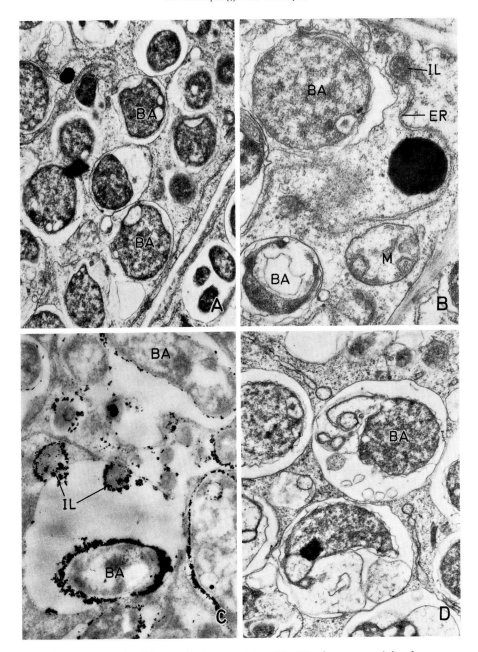

Plate 28. Necrosis of bacteroids in zone 4 (see Fig. 25) of a root module of pea.

A: The bacteroids *(BA)* are localized within membrane-bound vesicles. ×10,600. *B, C:* Involvement of primary lysosomes *(IL)* in the transformation of bacteroid containing vesicles into heterophagic vacuoles; primary lysosomes that are given off by the ER *(B* ×21,000) provide the vesicles with digestive enzymes (*C* Gomori reaction, ×31,200). *D:* Disintegrating bacteroids in heterophagic vacuoles. ×15,400. Courtesy of G. TRUCHET and PH. COULOMB.

and COULOMB 1973). It seems that the digestion of the bacteroids coincides with the senescence and eventual autolysis of the host cell. Another case of apparent heterophagy in a higher plant was recently reported by COULOMB (1973 b). She claims that in meristematic root cells of *Scoroconera hispanica* a phagocytosis-like process results in the formation of multivesicular bodies which contain polysaccharides. These bodies are eventually transfered to vacuoles. In this case heterophagy would be identical with the transfer of some sort of cell wall material from the external space into the vacuoles. Since the biochemical counterpart and the functional significance of this process are unknown, it cannot be decided whether it can be compared to the heterophagy of components of cartilage in the connective tissue of animal cells (see DINGLE 1969).

3.1.6. Autolysis and Cell Death

Senescence ends with the abolition of compartmentation. The chaotic appearence of the cytoplasm and the diffused distribution of acid phosphatase become evident when senescence has proceeded beyond a certain point (BERJAK and VILLIERS 1972). An analysis of events in the senescing corolla of *Ipomoea* has shown that prior to the disorganisation of the whole protoplast the vacuoles shrink and the cytoplasm loses its normally dense appearance (Plate 29; MATILE and WINKENBACH 1971). This morphological phenomenon suggests that severe changes in membrane permeability precede the final phase of cell ageing. It has in fact been noticed that in ageing leaf sections as well as in ripening fruits the space accessible for mannitol is progressively increased towards the end of senescence (SACHER 1967, 1973). Finally, autolysis is brought about by the rupturing of the tonoplast which results in the free mixing of vacuolar hydrolases with all the cytoplasmic constitutents. It is only after hydrolase compartmentation has been abolished that the progressive degradation of nuclei can be observed in disorganized mesophyll cells of *Ipomoea* petals (Fig. 26; Plate 29; MATILE and WINKENBACH 1971). The asynchronous behaviour of individual cells in this tissue is documented by the fact that autolysed cells can be seen throughout senescence. The decline of DNA (see Fig. 19 A) seems to indicate the progress of cell autolysis, since the bulk of the DNA is contained in the nuclei which are not digested until the tonoplast is ruptured.

Essentially comparable sequences of events in late senescence, especially regarding the rupturing of vacuolar membranes, have been discovered in a variety of objects (SHAW and MANOCHA 1965, BUTLER 1967, TREFFRY *et al.*

Plate 29. Autophagy and autolysis in mesophyll cells of the fading corolla of *Ipomoca tricolor.*

A: Intensified autophagic activity. Note the presence of mitochondria (*M'*) membranes and ribosomes in the cell sap. ×16,000. *B:* Shrinkage of the vacuole, inflation of mitochondria (arrows) and dilution of the cytoplasm. ×16,000. *C:* Rupture of the tonoplast. ×40,000. *D:* Partially degraded nucleus after rupture of the tonoplast. ×12,000. *E:* Chaotic appearance of an autolyzing cell. ×16,000.

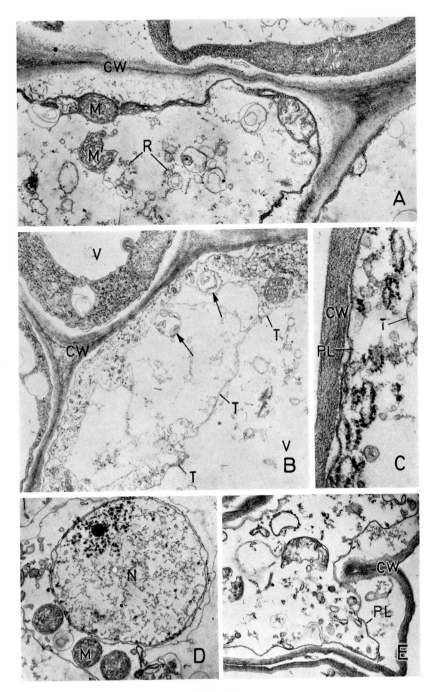

Plate 29.

1967, Ragetli *et al.* 1970, Berjak and Villiers 1970, Srivastava and Singh 1972, Wodzicki and Brown 1973, Berjak and Lawton 1973).

The herbicide paraquat induces various changes in the fine structure of flax cotyledons; if the cotyledons treated with paraquat are illuminated, tonoplast breakdown is evident after 6 hours and this is followed by progressive disorganisation of the cell contents (Harris and Dodge 1972). This finding not only illustrates the induction of autolysis by exogenous agents (toxins of microbial origin could also have similar effects), it also brings up the

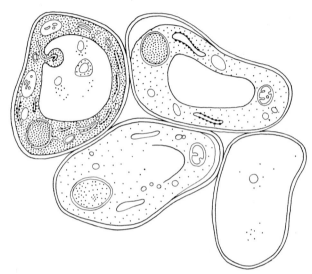

Fig. 26. Ageing and autolysis of mesophyll cells of the *Ipomoea corolla*. Adapted from Matile and Winkenbach (1971). Originally prepared for A. W. Robards: Dynamic Aspects of Plant Ultrastructure. Copyright 1974 by the McGraw-Hill Book Company (UK), Limited. Reproduced by permission.

question of how autolysis is triggered under normal conditions. It is very probable that the crucial event which irreversibly results in cell death is the rupturing of the tonoplast. However, it is unlikely that this is caused by the tonoplast itself. It seems rather that cessation of metabolic activity in the senescing cell, perhaps cessation of a continuous supply of building material or energy which is necessary to maintain the integrity of membranes, causes the decay of the tonoplast and, indirectly, the autolytic breakdown of the cytoplasm. Indeed, in the basidiomycete *Coprinus lagopus* the onset of autolysis in the gills coincides with the sharp decline of respiratory enzyme activity (Iten and Matile 1970). Another example may illustrate the dependency of hydrolase compartmentation on the availability of membrane constituents. Mutants of *Neurospora* requiring inositol can be induced to "commit suicide" if the conidia are allowed to germinate in the absence of exogenous inositol (Strauss 1958). Inositol-starvation in these mutants results in membrane damage which appears to be due to the liberation of proteinase from vesicles the membranes of which are comparatively rich in phosphatidyl

inositol (MATILE 1966). This phenomenon could provide a key for the understanding of ageing (SULLIVAN and DEBUSK 1973) and particularly for
autolysis. The breakdown of lysosomal compartmentation in the ageing cell
could be a consequence of membrane damage caused by the cessation of a
continuous supply of membrane constitutents synthesized in the cytoplasm.

3.2. Storage and Mobilization

The cyclic development of plants is characterized by periods of quiescence
which are preceded by the withdrawal of substances from the leaves and the
concomitant accumulation of reserves which form the nutritional basis for the
initial phase of growth in the next period of development. Storage tissues of
seeds, rhizomes, tubers and other storage organs are the sites of temporary
deposition of reserves.

The accumulation in cells of large quantities of carbon, nitrogen, and other
elements requires the formation of osmotically inactive or poorly active compounds. Macromolecular forms of reserves such as polysaccharides and
proteins or hydrophobic substances such as triglycerides meet this prerequisite.
They are deposited in plastids (starch), aleurone vacuoles (reserve proteins)
or spherosomes (neutral fat). In addition, macronutrients such as phosphate,
Ca, Mg, and K (phytate) are accumulated in aleurone vacuoles.

Many storage tissues have a limited life-span beyond the period of accumulation of reserves. In other words, the storage products which are laid
down intracellularly are mobilized within living cells. This is especially
obvious in cotyledons which develop into green leaves after the mobilization
of reserves is completed. The importance of compartmentation is particularly
evident in the case of the mobilization of reserve proteins through the action
of unspecific proteinases whereas the intracellular breakdown of starch in the
amyloplasts does not seem to require the involvement of the lytic compartment.

There are, however, storage tissues with cells which die off after the completion of reserve deposition. Mobilization of macromolecular storage products in such tissues evidently requires the production of hydrolases in a living
tissue of the seed and their release into the dead tissue. With regard to the
active tissue, mobilization proceeds extracellularly in these seeds. Examples
of dead storage tissues are given by certain endosperma, the most intensely
investigated being the starchy endosperm of the barley caryopsis.

3.2.1. Intracellular Digestion of Reserves

3.2.1.1. Accumulation and Mobilization in Aleurone Vacuoles

Aleurone grains of quiescent seeds contain large quantities of a few storage
proteins; in addition, these organelles are characterized by the presence of
proteinases (see p. 34). This appears to be significant as the degradation of
storage proteins is initiated immediately upon the soaking of seeds.

In the course of germination the dissolution of reserve protein is associated with
extensive inflation of aleurone grains which coalesce and assume the appear-

ance of vacuoles (Plate 30 *A*). The electron opaque masses of protein which characterize the aleurone grains of quiescent seeds gradually disappears. In large vacuoles of *Cucurbita maxima* cotyledons irregularly shaped bodies of undigested protein can be observed after 2–3 days of germination (Plates 30 *B*, 31 *B;* LOTT and VOLLMER 1973). The corroded appearance of the globoid crystals present in aleurone grains of this organ (Plate 31 *A*) suggests that

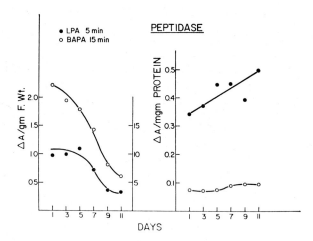

Fig. 27. Changes in peptidase activity in extracts of pea cotyledons following germination. The substrates were *LPA* (Leucine-p-nitroanilide) for amino-peptidase, *BAPA* (α-benzoyl-arginine-p-nitroanilide) for a trypsin type peptidase. From BEEVERS (1968).

phytic acid is also mobilized at an early stage of germination. Similar observations have been made in a number of germinating seeds. JONES (1969 b) has noticed that in the barley aleurone cells induced to mobilize reserves, vesicles which originate from the endoplasmic reticulum may fuse with aleurone vacuoles. This observation suggests that hydrolytic enzymes are newly synthesized in the course of germination and released into the digestive compartment.

Although it has been reported that the initial proteolytic activities are not increased in the course of germination (*e.g.*, in bean cotyledons: PUSZTAI and DUNCAN 1971), it seems that this may be true only in the case of exopeptidases (Fig. 27). Recently, YOMO and SRINIVASAN (1973) demonstrated that a

---

Plate 30. Mobilization of reserve protein in aleurone vacuoles.

A: Cross section of a cotyledon of a *Vicia faba* seed germinated for 7 days. In cells adjacent to the vascular bundle on the left, the aleurone grains have fused to form large vacuoles (*V*); the reserve protein has disappeared. Further away from the vascular bundle the degree of protein degradation in aleurone grains decreases. ×300. Courtesy of L. G. BRIARTY. *B:* Portion of a cotyledon mesophyll cell from a *Cucurbita maxima* seed germinated for 4 days at 31 °C. A central vacuole which contains irregularly shaped protein bodies (*PB*) has formed upon the coalescence of smaller vacuoles. ×18,500. Courtesy of J. N. A. LOTT.

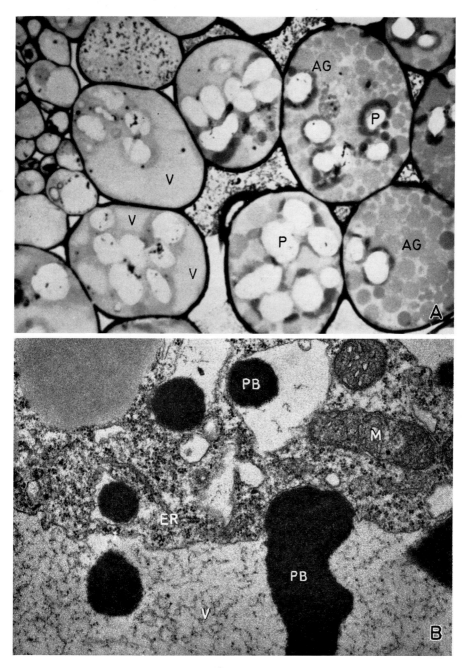

Plate 30.

considerable increase in endopeptidase activity occurs in cotyledons of germinating bean seeds. Similar findings in cotyledons of other species suggest that this may be a general phenomenon (Fig. 28; BEEVERS 1968, GARG and VIRUPAKSHA 1970, SZE and ASHTON 1971). The importance of the technique used for measuring the enzyme activities in homogenates appears from a recent study of CHRISPEELS and BOULTER (1974). When autodigestive proteo-

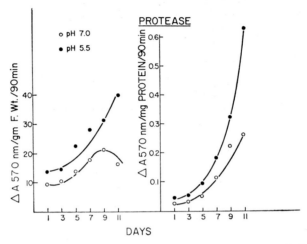

Fig. 28. Changes in proteolytic activity in extracts of pea cotyledons during germination (substrate: casein). From BEEVERS (1968).

lytic activities were measured in homogenates from cotyledons of germinating mung beans, a four to sixfold increase occurred during the third to fifth days of germination; true endopeptidase activity, assessed viscosimetrically with gelatine as the substrate, increased, however, ten to fifteenfold. The coincidence between the rapid decline of protein and the increase of gelatinolytic activity suggests that the appearence of endopeptidase may be a prerequisite for the rapid degradation of storage proteins.

SHAIN and MAYER (1968) were unable to detect incorporation of radioactive sulfur (supplied as sulfate) into the trypsin-like enzyme present in germinating lettuce seeds, yet this proteinase activity is greatly increased in the course of germination. This may lead to the conclusion that proteinases are activated, rather than synthesized *de novo*. However, results from

---

Plate 31. Changes in the fine structure of aleurone grains of *Cucurbita maxima* cotyledons during germination.

A: After 2 days of germination digestion of the globoid "crystal" (*GC*) has begun through the formation of internal pitted regions (arrows); proteinaceous matrix (*PM*), soft portion of globoid (*SG*). Mesophyll cell prefixed with glutaraldehyde and treated with 20% glycerol prior to freeze-etching. ×30,000. *B:* Portion of a spongy mesophyll cell from a cotyledon of a seed germinated for 3½ days. Irregularly shaped protein bodies (*PB*) are located in a large vacuole bounded by the tonoplast. ×6,050. Courtesy of J. N. A. LOTT.

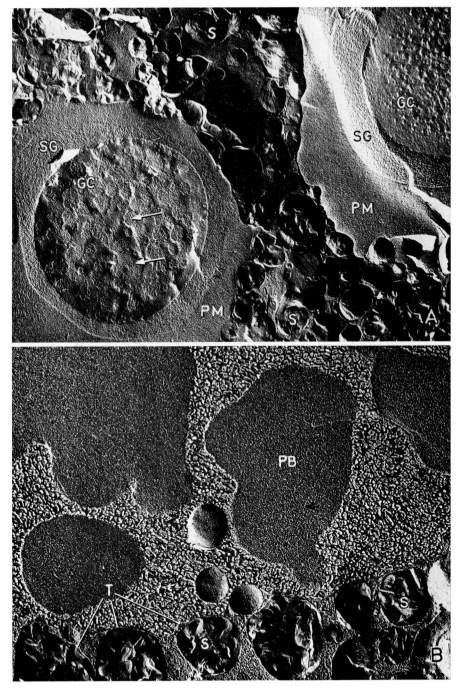

Plate 31.

radioactive labelling experiments are difficult to interpret since negative results may be caused by the preferential incorporation into protein of amino acids produced from storage proteins. The elucidation of facts is rendered more difficult by the presence in lettuce seeds of a specific inhibitor protein which inactivates the trypsin-like endopeptidase. In germinating lettuce seeds the gradual decrease of inhibitory activity (SHAIN and MAYER 1968) suggests that the increase in endopeptidase activity is at least partially due to activation. It is not justifiable, however, to conclude from these results, which were obtained with homogenates, that inhibitor proteins have a regulatory function in living cells. A prerequisite for this would be that the inhibitor protein and the corresponding proteinase be present in the same compartment. In pea seeds at least some of the trypsin inhibitor protein which is associated with the fraction of protein bodies isolated from cotyledons can be washed off (HOBDAY et al. 1973). Since this is the case also with other constituents of aleurone grains, this result does not necessarily indicate that the inhibitor protein is located outside the organelle; the aleurone grains are relatively fragile, especially if isolated from germinating seeds, so that constituents are possibly lost upon damage of the membrane.

Proteinase inhibitors are widely distributed in seeds (see RYAN 1973). It is tempting to speculate that they function as regulators of proteolysis in the germinating seeds. However, most of these inhibitors seem to be inactive toward the endogenous proteinases. This holds good, for instance, in the case of proteinase inhibitors of legume seeds which are accumulated in comparatively large amounts; in soybeans they represent about 6% of the total protein. These facts support the view that at least some of the inhibitor proteins are storage proteins without a regulatory function in the degradation of protein.

The deposition of storage proteins may not be restricted to seeds and vegetative reserve organs. The accumulation of proteinaceous deposits in vacuoles has also been observed in vegetative apices, floral apices and petals (Plate 32) of various species of *Solanaceae,* maize (SHUMWAY et al. 1972), *Nigella* (GREYSON and MITCHELL 1969), and *Rhododendron* (SCHNEIDER 1972) as well as in the inactive glands of Venus's flytrap (SCALA et al. 1968). Protein bodies present in vacuoles in tomato leaf cells (Plate 33) have been identified indirectly with a proteinase inhibitor (chymotrypsin inhibitor I protein) which, under certain conditions, is synthesized in comparatively large amounts (SHUMWAY et al. 1970). In detached tomato leaflets the synthesis of this protein is induced if illumination is provided (Fig. 29) and its accumulation is conclusively correlated with the appearance of electron dense clusters or globules in the vacuoles of mesophyll cells (Plate 33; RYAN and SHUMWAY 1971). Accumulation is reversible, hence, proteinases that are not inhibited by this storage protein (which interacts with a wide variety of proteinases) must be responsible for the mobilization. The function of chymotrypsin

---

Plate 32. Vacuolar protein bodies (*VP*) in apical and petal cells.

A: Vegetative apex of tobacco. ✕3,600. *B:* Floral apex of tomato. ✕3,240. *C:* Mature petal of tobacco. ✕3,570. Courtesy of L. K. SHUMWAY and C. A. RYAN.

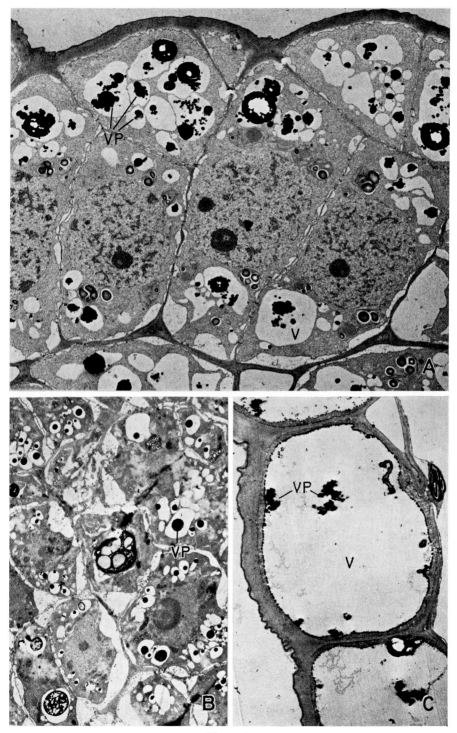

Plate 32.

inhibitor protein as a store of nitrogen in young leaflets of potatoes appears from a study of RYAN and HUISMAN (1967); upon the development of these organs the inhibitor is accumulated in an early period and later disappears (Fig. 30). Similarly, proteinase inhibitors are temporarily accumulated in young barley rootlets (KIRSI and MIKOLA 1971). Hence, protein storage and utilization in developing plants appears to be (at least in certain plants) very

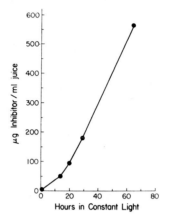

Fig. 29. Changes in proteinase inhibitor concentration with time after detachment of tomato leaves. The data shown are from the four halves of the leaflet pairs of one leaf. Electron micrographs of the same tissue are shown in Plate 33. From SHUMWAY *et al.* (1970).

complex; storage proteins present in potato tubers may be mobilized upon the induction of bud development, amino acids being transported into the buds and partially laid down there in the form of inhibitor protein, which is eventually mobilized as the leaflets grow out.

The temporary deposition of nitrogen may, however, be only one function of these proteinase inhibitor proteins. Another function is suggested by the observation that its synthesis in potato leaves is induced by Colorado potato beetles (GREEN and RYAN 1972). This seems to be an unspecific response to wounding since mechanical damage has the same effect of inducing the rapid accumulation of inhibitor protein in the wounded leaves as well as in the adjacent, undamaged leaves. GREEN and RYAN (1972) suggested that the inhibitor protein impairs the digestion of protein in the intestinal tract of the insect and, therefore, represents a defense mechanism.

Plate 33. Vacuolar protein bodies in tomato leaf cells: induction of chymotrypsin inhibitor protein I in detached leaves (compared with Fig. 29).

A: Mature spongy parenchyma cell of a tomato leaflet at the time of detachment from plant. The leaf has no inhibitor protein I and the vacuoles typically lack electron-dense material. ×3,330. B: After 14 hours in inducing conditions the leaflets contain 50 μg/ml of protein inhibitor I: Note the presence of vacuolar protein bodies (*VP*) in the central vacuole and in small vesicles (arrow) in the cytoplasm of a mature palisade parenchyma cell. ×3,330. *C:* After 64 hours under inducing conditions 565 μg of inhibitor protein are present per ml of leaf juice and large protein bodies are present in vacuoles. ×5,500.
Courtesy of L. K. SHUMWAY and C. A. RYAN.

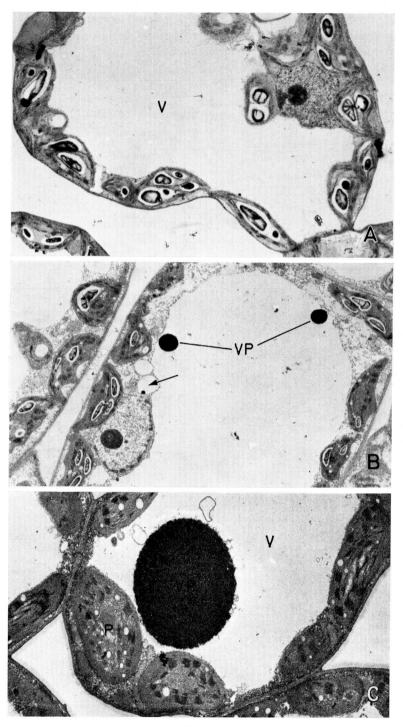

Plate 33.

It appears then that proteinase inhibitor proteins may have several functions, as regulators of proteolytic activity, as storage proteins, and as "toxins" for herbivorous insects. In all these functions the lytic compartment, that is vacuoles or aleurone grains, seem to be involved in their accumulation and in their eventual degradation.

Turning back to aleurone vacuoles a word should be said about storage products others than proteins. With regard to the involvement of hydrolases in the process of mobilization, phytic acid is of particular interest. Its degradation in the rice endosperm is correlated with a marked increase in phytase

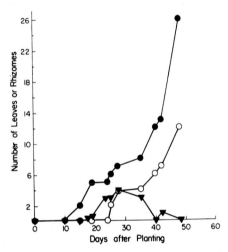

Fig. 30. Proteinase inhibitor protein present in the young potato plant. A plot of the total number of leaflets (●) and rhizomes (O); the number of leaflets showing the presence of inhibitor protein (▲) has been determined in an immunological double diffusion assay. From RYAN and HUISMAN (1967).

activity (MUKHERJI et al. 1971). The hydrolysis of phytate not only mobilizes phosphate but also the metal ions potassium, magnesium and calcium which are present in the salts of inositol hexaphosphate (EASTWOOD and LAIDMAN 1971). An interesting difference between the action of phytase in solution and in germinating wheat was reported by MATHESON and STROTHER (1969); in situ the mobilization of phytic acid does not involve the formation of myo-inositol phosphate esters with fewer than six phosphate groups, whereas in vitro the lower esters are formed before completion of hydrolysis. This could indicate that in the globoid of aleurone vacuoles phytase is bound to an individual molecule of phytic acid until all the six ester bonds are cleaved. Incidently, myo-inositol, one of the end products of phytate mobilization, is an important reserve substance, in that it is incorporated into inositol-phospholipids in the germinating wheat seed (MATHESON and STROTHER 1969).

It was suggested that phytate represents a pool of high energy phosphate which is used for the generation of ATP (MORTON and RAISON 1963); this possibility has, however, been disproved (WILLIAMS 1970). Phosphate seems to be present in phytic acid only in the monoester form. This could, in turn, lead to the speculation that not all the apparent phytase activities found are specific; unspecific acid phosphatase may be active also towards phytic acid, and observations by Poux (1963 b, 1965) on the concentration of acid phosphatase around the globoids of aleurone grains seem to support this view.

3.2.1.2. Spherosomes and Neutral Fat

Problems bearing on the interpretation of biochemical properties of spherosomes have already been touched upon above. Whether or not these organelles are considered as constituents of the lytic cell compartment, it is well established that they are the depositories for triglyceride in cells of higher plants. This seems to apply to the relatively large oil droplets in oleaginous seed tissues such as peanut cotyledons (JACKS et al. 1967) or tobacco endosperm (MATILE and SPICHIGER 1968, SPICHIGER 1969), as well as to the smaller spherosomes in leaves (YATSU et al. 1971) and cotyledons of legumes (ALLEN et al. 1971).

In seeds, triglycerides accumulated in spherosomes represent an osmotically inactive reserve of carbon which, upon germination, is transformed into sugar. Gluconeogenesis involves the concurrence of *glyoxysomes*. In these membrane-bound cell organelles the enzymes of the β-oxidation of fatty acids and of the glyoxylic acid cycle are localized (BEEVERS 1969). The metabolic cooperation between spherosomes and glyoxysomes in germinating seeds is clearly evident in electron micrographs such as in Plate 34 A showing glyoxysomes surrounded by oil droplets (CHING 1970, MOLLENHAUER and TOTTEN 1970, VIGIL 1970). The factor which possibly determines the flow rate of fatty acids from spherosomes to glyoxysomes is lipase; it hydrolyzes the lipophilic neutral fat molecules to hydrophilic fatty acids which are then released from the oil droplets. As already summarized above (see p. 39) the information on lipase localization in spherosomes is contradictory and spherosomes may, in fact, be biochemically heterogeneous. In any case, the presence of lipase in spherosomes of the *Ricinus* endosperm has been demonstrated (ORY et al. 1968) and the association of spherosomes with glyoxysomes has been observed in the same tissue (VIGIL 1970). In the tobacco endosperm the lipid stored in spherosomes is almost completely consumed after four days of germination. An interesting feature of lipid composition in spherosomes isolated at intervals from germinating seeds is the constant proportion of tri-, di-, and monoglycerides and fatty acids; at all stages of lipid mobilization about 86–87% of the spherosomal lipid is triglyceride whereas the relative content of the endosperm in free fatty acids increases gradually (SPICHIGER 1969). These findings suggest that lipase is (superficially?) associated with the spherosomes and gives rise to the release of free fatty acids. Indeed, incubations of isolated tobacco spherosomes have resulted in the saponification of half of the original triglycerides within 8 hours. It is remarkable that this was done with lipid globules isolated from dormant seeds (SPICHIGER 1969); hence, lipase seems to be synthesized and integrated into spherosomes upon the development of the tobacco endosperm so that lipolysis will come into action immediately after soaking of the seeds.

In the castor bean endosperm an alkaline lipase which is particularly active during the third to fifth day of germination and which hydrolyzes only monosubstituted glycerols is associated with the glyoxisomal membrane (Muto and Beevers 1974). It is likely that the saponification of triglycerides by the action of the spherosomal acid lipase yields monoglycerides (apart from free fatty acids) which are ultimately hydrolyzed upon the movement into the glyoxisomal compartment.

There are, however, species of fat storing seeds the spherosomes of which seem to be completely devoid of lipase. According to Jacks et al. (1967) practically none of the lipase activity seen in germinating peanut cotyledons is associated with spherosomes; the greatest activity is associated with a sub-cellular fraction which sediments at $\times 15,000$ g. This fraction, although designated as the mitochondrial fraction, most probably contains the aleurone vacuoles. To date nothing is known about the location of lipase in these organelles, but morphological observations reported by several authors suggest that spherosomes may be autophagocytized by vacuoles. Poux (1963 b, 1965) particularly pointed out that spherosomes may be present in aleurone vacuoles. In addition, the engulfment of spherosomes by tonoplasts (Plate 34 B, C) and the obvious presence of lipid globules in vacuoles has been observed repeatedly in tissues other than reserve tissues of seeds (Lowry et al. 1967, Ruinen et al. 1968, Mittelheuser and Van Stevenink 1971, Schwarzenbach 1971, Maitra and Deepesh 1972).

As already pointed out above, lipase seems to be present in the quiescent seed, at least in certain species. A considerable increase in lipolytic activity concomitant with the mobilization of triglycerides has been found in the germinating seed of Douglas fir (Ching 1968). In the wheat grain, lipase activity is very low at the beginning of germination and rises to a high level between the second and sixth day of germination (Tavener and Laidman 1972). A similar increase in lipase activity has been observed during the culture of isolated apple embryos (Smolenska and Lewak 1974).

Not all types of spherosomes seem to function as storage sites for neutral fat which is eventually transformed into sugar. The spherosomes of baker's yeast are biochemically distinct from spherosomes in higher plants; according to Schaffner (1974) they contain sterolesters and phospholipids apart from triglycerides and, therefore, appear to function as storage sites for membrane lipids. A similar function might be attributed to spherosomes that are lined up adjacent to the plasmalemma of the cells of pea and bean cotyledons;

Plate 34. Mobilization of triglycerides in spherosomes.

A: Germinating seed of *Ricinus communis:* section of an endosperm cell with spherosomes and numerous microbodies (*MB*) (glyoxisomes). $\times 7,400$. The intimate association between glyoxisomes and spherosomes is shown in the inset where eight microbodies are apposed to a single spherosome. $\times 15,700$. Courtesy of E. Vigil. *B:* Engulfment of a large lipid droplet by a vacuole in *Saccharomyces cerevisiae.* Ultrathin frozen section; cytochemical demonstration of alkaline phosphatase (arrows). $\times$ ca. 30,000. Courtesy of H. Bauer. *C:* Engulfment of spherosomes by vacuoles in a cell of the embryo of a germinating *Ricinus* seed. $\times 8,500$. Courtesy of A. Schwarzenbach.

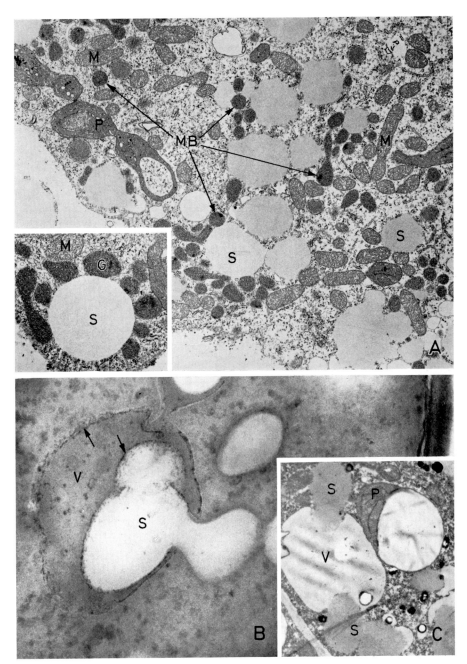

Plate 34.

upon germination these composite, thread-like lipid vesicles are converted into saccules (Plate 14 E), evidently because their contents have to be evacuated (MOLLENHAUER and TOTTEN 1971). In contrast to the simple spherosomes which are also present in these cells, the composite vesicles are characterized by a comparatively high content in phospholipids (ALLEN et al. 1971).

3.2.2. Extracellular Digestion of Reserves

Only a few examples of reserve mobilization involving the secretion of hydrolases shall be mentioned here. The phenomenon of extracellular digestion will be discussed in more detail in connection with the regulation of lytic processes (section 4.1.). In particular, the mobilization of reserves accumulated in the dead cells of the starchy endosperm of barley seeds, which is associated with the secretory activity of the aleurone layer (review by YOMO and VARNER 1971), will be discussed from the point of view of control mechanisms.

In contrast to the accumulation of intracellular reserves in the starchy endosperm of barley, the development of the endosperm in some other seeds is characterized by the deposition of carbohydrate in the form of voluminous secondary cell walls. These walls consist of hemicellulosic polysaccharides. In the endosperm of *Phoenix dactylifera* seeds the hard cell walls consist predominantly of mannan. By the end of endosperm development the space of the cytoplasm is reduced to a small lumen. In certain leguminous seeds the endosperms contain large amounts of mucilaginous galactomannan which is also deposited extracellularly (REID and MEIER 1973 a).

Since the cells of these reserve tissues die off at the end of seed development, the mobilization of polysaccharides requires the secretory activity of an adjacent living tissue. In the date seed the haustorium of the embryo has a dual function in that it not only secretes the appropriate hydrolases but also absorbs and transforms the sugars produced upon the progressive digestion of the mannan (KEUSCH 1968). In the legume *Trigonella foenum-graecum* the breakdown of galactomannan in the dead moiety of the endosperm (Plate 35; REID 1971) is associated with the secretion of α-galactosidase, β-mannosidase and probably endo-β-mannanase by the aleurone layer (REID and MEIER 1973 b).

Plate 35. Mobilization of galactomannan in the endosperm of germinating *Trigonella foenum-graecum* seeds.

A: Cross section of the outer part of a seed before mobilization of the galactomannan, showing the seed coat (*SC*), a small portion of a cotyledon (*C*) and between these two the endosperm. The aleurone layer (*A*) is the outer layer of the endosperm, the rest being composed of large cells which have thin primary walls and are completely filled with dark stained galactomannan (*GM*). ×600. *B:* During mobilization the galactomannan is gradually dissolved, the dissolution zone beginning at the aleurone layer. The primary walls are not dissolved. ×800. Courtesy of J. S. G. REID.

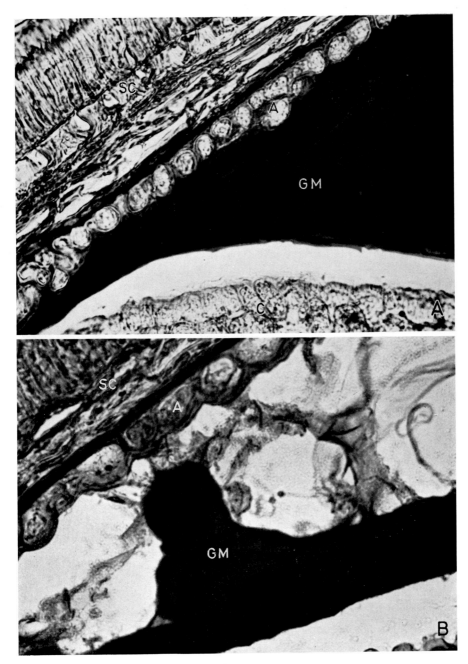

Plate 35.

3.3. Cell Wall Lysis in Development

Changes in cell wall structure caused by secreted lytic enzymes have already been mentioned in connection with the extracellular release of macromolecules (see section 1.4.): β-glucanase secretion in aleurone cells of barley appears to cause the perforation of the walls and thus renders them "permeable" for the α-amylase and other hydrolases subsequently released. This example is outstanding because both aspects of the phenomenon, the metabolic and the cytological, were considered in the study of TAIZ and JONES (1970). In many other cases of obvious cell wall degradation the available information is much less complete. The reader will certainly recall the plant cytology course in which the callose plugs in the resting sieve tubes of *Vitis* stems were demonstrated; these sieve tubes are reactivated in spring, which involves the removal of the callose deposits in the sieve pores. This is evidently a cell wall lytic process and the existence in *Vitis* of the corresponding enzyme, β-1,3-glucan hydrolase, which is probably responsible for the unplugging of sieve plates, has been established (CLARKE and STONE 1962). However, the cytological phenomena associated with the local operation of glucanase in the sieve plate region have not yet been investigated.

Examples of *obvious* cell wall lysis are abscission of senesced leaves and ripe fruits, or cell fusion processes such as in zygote formation which can take place only after the cell walls have been at least locally lysed by the action of corresponding hydrolases (*e.g.*, SASSEN 1965, CLAES 1971). In addition to these obvious lytic events in cell walls, increasing evidence supports the view that *hidden* lysis represents one aspect of a conspicuous dynamism of plant cell walls. In fact, the rigid cell walls appear to hinder the cell extension which plays a prominent role in plant growth. It seems that cell wall lytic enzymes are deeply involved in bringing about the necessary plasticity of cell walls which undergo deformation.

3.3.1. Cell Wall Metabolism in Growing Cells

Metabolic turnover is not only a phenomenon associated with cytoplasmic cell constituents, it has also been demonstrated that it occurs in cell walls. The metabolic lability of cell wall polysaccharides is reflected by the profound changes of sugar composition that are associated with growth. In meristematic tissues of seedlings sugars such as rhamnose, arabinose, mannose and galactose account for 70% of the wall mass, whereas in the fully expanded tissue these constituents of hemicelluloses drop to only 10–20% (NEVINS *et al.* 1968). Similar changes are associated with auxin-induced growth of *Avena* coleoptiles (LOESCHER and NEVINS 1972). Radioactive labelling experiments with growing organs have indeed revealed metabolic lability of cell wall constituents. Substantial fractions of the hemicelluloses are turned over and the extent of this is correlated with growth; in contrast, cellulose does not appear to be subjected to degradation (MACLACHLAN and JOUNG 1962, MATCHETT and NANCE 1962, KATZ and ORDIN 1967 a, WADA *et al.* 1968). The significance of a partial degradation of cell wall polysaccharides is

vividly expressed by the term "cell wall loosening". It has been known for
a long time that cell walls of growing organs are characterized by greater
plasticity but only in the past few years has this phenomenon been associated
with cell wall lytic processes. An original way of demonstrating the involvement of specific hydrolases in wall loosening and induction of growth is the
employment of exogenous, purified enzymes. MASUDA and his colleagues used
fungal glucanases to evoke the elongation of coleoptile segments (TANIMOTO
and MASUDA 1968, MASUDA and SATOMURA 1970) whilst ROGGEN and
STANLEY (1969) demonstrated that exogenously applied pectinase, β-1,3- and

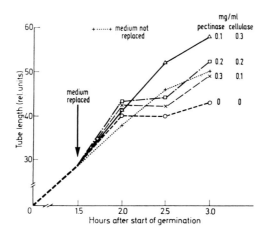

Fig. 31. The effect of exogenous pectinase and cellulase on the growth of *Pyrus communis*
pollen tubes *in vitro*. From ROGGEN and STANLEY (1969).

β-1,4-glucanase increase the rate of extension in pollen tubes *in vitro* considerably (Fig. 31). In contrast, RUESINK (1969) was not albe to find any
significant effect of purified *Myrothecium* cellulase on the elongation of *Avena*
coleoptiles. This contradictory result indicates that cellulose degradation may
not be involved in wall loosening, which is also suggested by the absence of
turnover in this polysaccharide. However, DATKO and MACLACHLAN (1968)
observed a conspicuous increase in cellulase activity in decapitated pea apices
treated with auxin; the enzyme was found to be associated with isolated cell
walls as well as with microsomes. In these experiments cellulase activity was
assessed with carboxymethylcellulose as the substrate. This viscosimetric assay
yields information about the cleavage of β-1,4-glucosidic bonds in an endo-
fashion which does not necessarily imply that the enzyme is also active on
native cellulose. Another explanation for the absence of effect of exogenous
cellulase on elongation growth would be that this enzyme induces wall
loosening only in conjunction with other hydrolases.

   Whether or not cellulase is involved in wall loosening, there is little doubt
that growth requires the activity of other, endogenous hydrolases. According
to NEVINS (1970) β-glucosidase activity, which is firmly bound to cell walls
of bean hypocotyls, peaks immediately preceding the logarithmic growth

phase and this is closely correlated with a marked decrease in wall glucose (Fig. 32). Other hydrolases such as β-galactosidase and β-xylosidase decline as a function of tissue age. It will be seen later when the control of lytic processes in cell wall metabolism is considered (see p. 139) that the levels of hydrolases measured in homogenates do not necessarily reflect their activities *in vivo;* growth seems to be associated with changes of pH in the mural space and studies of hydrolase involvement should take this into account. It can be assumed that in cells cultured in suspension the pH of the mural space corresponds to the pH in the walls, therefore the finding of Keegstra and

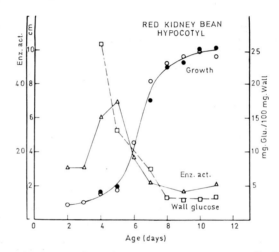

Fig. 32. Relationship between β-glucosidase activity (Enz. Act.), wall glucose and growth of the bean hypocotyl. The enzyme activity is expressed as μmoles p-nitrophenol released from p-nitrophenyl-β-glucoside/hr × g fresh weight. From Nevins (1970).

Albersheim (1970) that the activity of a number of external glycosidases in cells of *Acer pseudoplatanus* cultured in suspension is intensified as the cells go through a period of growth, appears to be significant. These enzymes are indeed capable of degrading isolated cell walls to a limited extent (Keegstra and Albersheim 1970). A similar partial autolysis of isolated cell walls has been reported also by Lee *et al.* (1967). The extent of this effect seems to be correlated with growth, as demonstrated by the intensified autolysis upon the treatment of *Avena* coleoptile segments with auxin (Katz and Ordin 1967 b).

In order to avoid the removal of hydrolases that are only loosely associated with walls, glycerol has been used as a nonaqueous medium for cell wall isolation. Indeed, when incubating cell wall fragments of corn coleoptiles isolated in glycerol the original weight is considerably reduced by autolysis; the analysis of the solubilized polysaccharide has shown that a polymer composed of β-1,3- and β-1,4-linked glucose residues is released from the walls (Kivilaan *et al.* 1971). It thus appears, that specific wall components such as this lichenan-like hemicellulose may impart structural rigidity to the primary cell wall. The lysis of these polysaccharides which takes place during cell elongation may be responsible for the necessary wall plasticity.

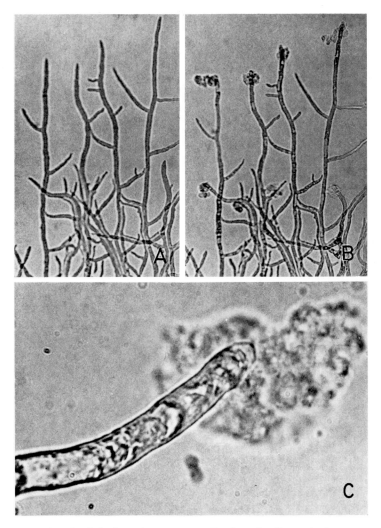

Plate 36. Bursting hyphal tips of *Mucor rouxii* after flooding colonies grown on full strength agar with distilled water.

A: Micrograph taken immediately after addition of water. *B:* One minute later. *C:* Detail of a burst tip. *A, B* ×115; *C* ×1,000. Courtesy of S. Bartnicki-Garcia.

The mere correlation of partial cell wall lysis with hydrolase activity is not fully satisfactory because growth is an organized process and, hence, the lytic activity must also be organized. It would seem that extension growth of axial tissues requires wall loosening (and of course, concomitant polysaccharide synthesis) in the side rather than in the end walls. The necessity of a highly organized release of cell wall lytic hydrolases is particularly evident in cells showing tip growth such as root hairs and pollen tubes. Nothing is known about the vectorial release of these enzymes into the mural space in cells of

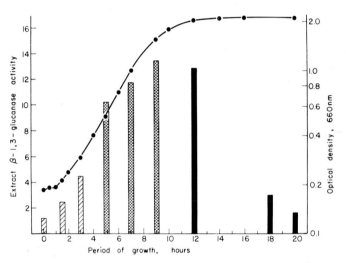

Fig. 33. Activity of β-glucan hydrolase during growth of *Schizosaccharomyces pombe*. Enzyme activities were measured on three separate cultures, as indicated in the histograms. Growth (●————●) was followed by measuring the optical density of cell suspensions at appropriate dilutions. From BARRAS (1972).

higher plants. However, in fungi, the growth pattern of which can be compared with that of pollen tubes or root hairs, some evidence favouring an oriented vesicular secretion of cell wall lytic enzymes is available. It is interesting to note that filamentous fungi show a bursting tendency in the tip (BARTNICKI-GARCIA and LIPPMANN 1972). The same phenomenon has been observed in pollen tubes (MATCHETT and NANCE 1962). If *Mucor rouxii* is transferred from full strength agar to distilled water the hyphal tips tend to burst (Plate 36). As suggested by BARTNICKI-GARCIA and LIPPMAN (1972) this can be taken as "presumptive evidence for a delicate balance between wall synthesis and wall lysis in apical growth". The concentration of β-glucosidase, a possible mural enzyme, in the hyphal tip of *Aspergillus oryzae* (REISS 1969) lends support to this view (Plate 4 B). JOHNSON (1968) attempted to demonstrate the dissolution of glucan in yeast cell walls by indirect means. He impaired the synthesis of glucan in the presence of the glucose analogue 2-deoxyglucose; as a result of this treatment cells tended to lyse at budding sites which led to the proposition that "synthesis of glucan in yeast walls occurs by insertion of glucose into glucan chains at cuts made by glucanases". This interpretation of the effect of 2-deoxyglucose is apparently not quite correct since it has been shown that this compound is metabolized and even incorporated into cell wall polysaccharides; this seems to alter the wall composition making the cells more susceptible to lysis (BIELY et al. 1971). Nevertheless, it is most probably true that glucanases play an important role in the budding process. Certainly yeast cells do contain glucanases associated with the cell wall. This is illustrated by the fact that spheroplasts of *Saccharomyces cerevisiae* secrete β-glucanase into the medium

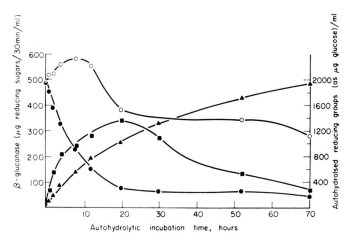

Fig. 34. Autohydrolysis of a crude cell wall preparation from *Schizosaccharomyces pombe*. The course of hydrolysis was followed by measuring the release of reducing sugars (▲). Samples were removed at intervals and assays for β-glucan hydrolase were made (O). In the course of autohydrolysis glucanase gradually disappears in the sedimentable wall fragments (●) and appears in the supernatant fraction of the cell wall preparation (■). From BARRAS (1972).

(BIELY et al. 1972). In *Schizosaccharomyces pombe* the activity of a β-1,3-glucan hydrolase associated with the cell wall is conspicuously correlated with growth (BARRAS 1972). As can be seen in Fig. 33 the total endoglucanase activity reaches its maximum during exponential growth and decreases sharply as the culture becomes stationary. This enzyme is not merely present, it is also capable of autohydrolysing isolated cell walls (Fig. 34).

Cyclical changes of glucanase and mannanase released into the medium of synchronously dividing cultures of brewer's yeast suggest that these enzymes function in the budding process: activity peaks coincide with the budding phase of the cell cycle (MADDOX and HOUGH 1971). A stepwise increase of laminarinase activity which precedes the extrusion of buds has also been observed in a synchronously budding population of baker's yeast (Fig. 35, CORTAT et al. 1972). Moreover, the release of this hydrolase into the mural space of budding *Saccharomyces* coincides with the budding phase (MATILE 1973). Further evidence for the involvement of cell wall degrading enzymes in the expansion of yeast cells stems from the work of SHIMODA and

Yanagishima (1971) on, surprisingly, auxin-induced volume increases which are correlated with increased glucanase activity.

In the budding process of yeasts the need for a localized improvement of cell wall plasticity is particularly obvious. This is achieved by a concentrated release of hydrolases such as glucanase. A morphological basis for such an oriented secretion is explained by several reports on the localized release of

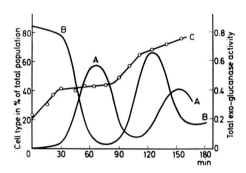

Fig. 35. Exo-glucanase activity in a synchronously budding population of *Saccharomyces cerevisiae*. *A* late budding cells; *B* initial budding cells; *C* exo-glucanase (laminarinase). Note the stepwise increase of total glucanase activity preceding the onset of budding. From Cortat *et al.* (1972).

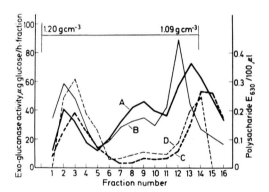

Fig. 36. Distribution density gradients of urografin of particulate exo-glucanase activity and polysaccharide (mannan) from growing and stationary phase cells of *Saccharomyces cerevisiae*. *A* glucanase activity; *B* mannan from exponentially growing cells. *C* glucanase activity; *D* mannan from stationary phase cells. Note the absence of particles of intermediate density in the non-growing cells. From Cortat *et al.* (1972).

vesicles in the region of growing buds (*e.g.*, Moor 1967, Sentadreu and Northcote 1969, McCully and Bracker 1972). An example of this is shown in Plate 18. These vesicles have been isolated from growing baker's yeast cells using glucanase activity as a marker in density gradient centrifugation (Cortat *et al.* 1972). The enzyme profiles obtained in gradients from growing and stationary cells, respectively, depicted in Fig. 36, demonstrate that one class of subcellular particles is present only in homogenates from budding cells. This material consists of vesicles with similar dimensions as the ER-derived vesicles observed by MOOR (1967) at the sites of budding. They show the glucanase activities required for glucan degradation and induction of wall plasticity; strong morphological evidence suggests that they are indeed released into the wall.

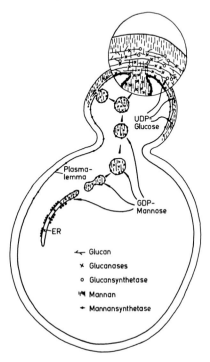

Fig. 37. Diagram showing the involvement of secretory vesicles in cell wall metabolism of budding yeast cells. From MATILE (1973).

From the above facts cell wall metabolism in the budding yeast cell can be depicted as in Fig. 37. It will be noted that the secretory "budding vesicles" contain not only glucanase but also mannan, a cell wall polysaccharide of yeast. Moreover, these vesicles contain mannan-synthetase and, therefore, closely resemble the Golgi vesicles in cells of higher plants (CORTAT et al. 1973). Hence, the original idea that expansion of cell walls in budding yeast cells is caused by local lysis and concomitant incorporation of wall polysaccharides appears to be true. The force which is responsible for the stretching of the locally loosened cell walls is, of course, the turgor. When the turgor is reduced by exposing yeast cells to osmotically active solutes, growth is immediately stopped and is not resumed until the turgor in the cells has been adequately adjusted.

Secretory vesicles which line developing walls have been observed in filamentous fungi (HEATH et al. 1971). They are particularly conspicuous

in the apical tips of various fungi (*e.g.*, GIRBARDT 1969, GROVE *et al.*
1970, GROVE and BRACKER 1970, COOK 1972) as can be seen in Plate 37 *A*.
The apical vesicles are considered to be secretory vesicles similar to
Golgi vesicles which fuse with the plasma membrane. This fusion is associated with the release of cell wall polysaccharides which eventually are integrated into the macromolecular complex of the wall. In addition they may,
however, be involved in the maintenance of wall plasticity if they contain
appropriate hydrolases. This function was suggested by COOK (1972) who
observed two classes of vesicles involved in the development of conidiophores
and spores of *Oedocephalum roseum;* the smaller type of vesicle seems to be
responsible for hydrolase secretion since its presence is notably associated with
cell extension. Indeed, the lysis of spore walls upon germination (Plate 37 *B*)
and extrusion of germ tubes (*e.g.*, LOWRY and SUSSMAN 1968, HESS and
WEBER 1973) seems to involve vesicles that collect along the plasmalemma in
the region of the ruptured spore wall (BRACKER 1971).

Evidently, the observations of localized vesicular secretion in budding and
in apical growth can tell us nothing about the causes of vectorial release of
polysaccharides and/or cell wall lytic enzymes. The demonstration of the
involvement of a distinct principle causing an organized secretion would
certainly be attractive, although, as is usual in biological research, the next
question would concern the principle which is responsible for the existence
and organization of this principle. Nevertheless, it is noteworthy that preliminary experiments with the fungal metabolite cytochalasin B (which interferes
with microfilaments) suggest the involvement of microfilaments (responsible
for the oriented release of vesicles?) in apical growth of fungi (BETINA *et al.*
1972, OLIVER 1973) as well as of pollen tubes (HERTH *et al.* 1972). According
to THOMAS *et al.* (1974) the secretion of cellulase, which in *Achlya* is associated with the branching of hyphae (see p. 56, 118), is indeed inhibited in the
presence of cytochalasin.

Finally, to round off this section, a couple of well documented developmental processes in which the lysis of cell wall components is evident, should be
mentioned. One of them was presented by WESSELS (1966) who was able to
correlate the formation of the pileus in *Schizophyllum commune* with the
activity of a glucanase which cleaves an alkali-insoluble glucan in this mush-
room. Using this glucan for measuring enzyme activity, glucanase began to
increase at the onset of the formation of the fruiting body. As can be seen

---

Plate 37.

A: Hyphal tip of *Armillaria mellea* showing apical cluster of vesicles. The irregular profile
of the plasmalemma in the apex suggests vesicular fusion with the plasmalemma. ×30,000.
Courtesy of S. N. GROVE and C. E. BRACKER. *B: Tilletia controversa* teliospore during
germination. All of the wall layers, except a small amount of the outermost wall layer,
have been dissolved. The wall of developing germ tube is marked by the arrow. ×15,600.
Courtesy of W. M. HESS. *C:* Infection of an onion root cell by *Pyrenochaeta terrestris.*
Note the beginning dissolution of the host cell wall. ×20,500. Courtesy of W. M. HESS.

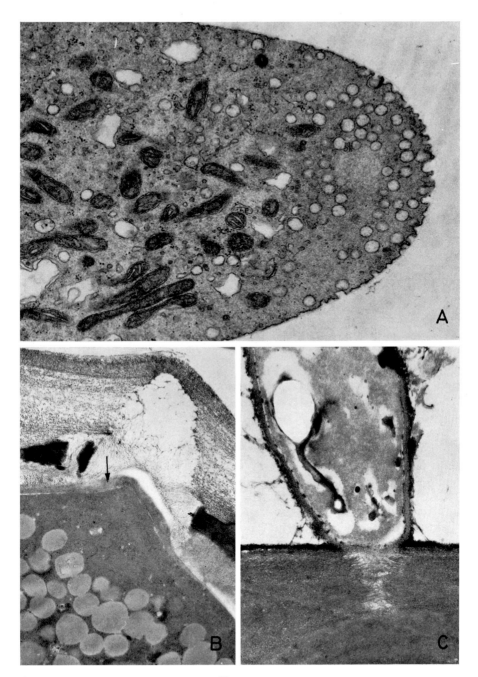

Plate 37.

in Fig. 38 this enzyme is released into the medium; a particularly interesting result is, however, that glucanase is also present in a particulate cell fraction which indicates the possibility of a vesicular type of secretion. Further evidence favouring the involvement of glucanase in the formation of the pileus was obtained with a morphological mutant of *Schizophyllum* with abnormal fruiting bodies which seem to be caused by a glucan that is less susceptible to enzymatic attack than the normal one (WESSELS 1966).

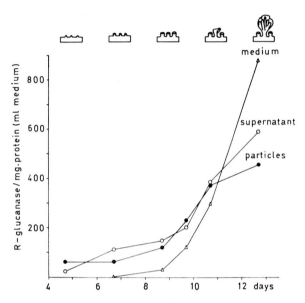

Fig. 38. Correlation between glucanase activity and pileus formation in *Schizophyllum commune*. The schematic drawings at the top represent the various stages of development. The glucanase activity were assessed using an alkali-insoluble cell wall glucan (R-glucan) of the fungus. Glucanase released into the culture medium is given per ml of medium; the intracellular glucanase present in cell-free extracts is expressed as specific activities in particulate and supernatant fraction respectively. From WESSELS (1966).

In the primitive fungus *Achlya*, which has cellulosic cell walls, the branching of hyphae, which can be hormonally or nutritionally induced, is accompanied by a conspicuous rise in cellulase activity (THOMAS *et al.* 1967). Inhibitor studies suggest that protein synthesis is involved in the induction of branching.

Cell wall lysis which is probably associated with cell autolysis has been studied in *Coprinus lagopus*. The fruiting body of this basidiomycete is dissolved shortly after its formation is completed (ITEN and MATILE 1970). Chitinases which seem to be responsible for the degradation of chitin in the cell walls are formed shortly before the beginning of autolysis of mature fruiting bodies and concomitant spore release. These enzymes are initially localized together with other hydrolases in vacuoles in the cells. Their release into the walls is associated with the induction of autolysis at the edge of the gills. Autolysis results in the formation of a liquor which is sucked by capillary forces into the intact regions of the gills where it induces cell lysis.

Thus, autolysis of the pileus appears to be an autocatalytic process: a zone of autolysis, at first evident at the edges of the gills adjacent to the stipe, rapidly moves to the periphery of the pileus reaching the edge within a period of only 5 hours (Figs. 39 and 40).

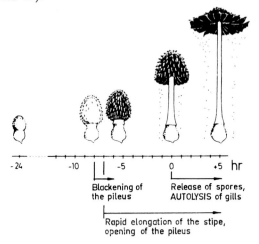

Fig. 39. Morphogenesis of fruiting bodies in *Coprinus lagopus*. The onset of spore release coincides with beginning of autolysis of gills. From ITEN (1969).

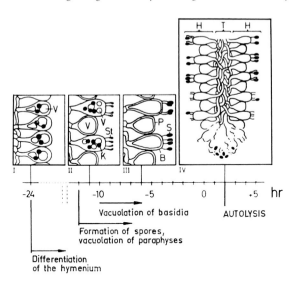

Fig. 40. Development of gills in the fruiting body of *Coprinus lagopus*. *I* undifferentiated hymenium; *II* beginning of sporogenesis; *III* sporogenesis completed; *IV* onset of autolysis and spore release. From ITEN (1969).

3.3.2. Senescence, Fruit-Ripening and Abscission

A substantial proportion of the dry mass of leaves is cell wall polysaccharide. That these carbohydrates can be used as a source of carbon by soil microorganisms is interesting from an ecological point of view, but the waste of these structural polysaccharides when the leaves fall contradicts the conception of an economical plant metabolism. Up to now comparatively little is known about the corresponding situation in senescing leaves. An early report by GÄUMANN (1935) contains some evidence that in the case of *Fagus* cell wall polysaccharides are partially recovered from the leaves before abscission. Further investigations on the metabolism of the ephemeral flowers of *Ipomoea tricolor* resulted in the surprising discovery that substantial proportions of

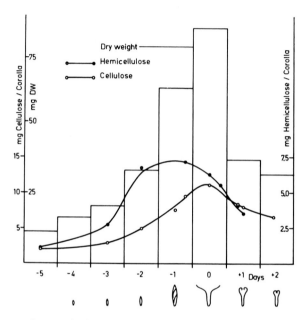

Fig. 41. Changes in dry weight, hemicellulose and cellulose contents of the *Ipomoea tricolor* corolla in the course of development. The sketches below the time axis indicate the shape of the corolla. After WIEMKEN-GEHRIG *et al.* (1974).

both cellulosic and hemicellulosic polysaccharides are degraded in the course of senescence (Fig. 41; WIEMKEN-GEHRIG *et al.* 1974). In this plant the partial degradation of the cell walls followed by the withdrawal of sugars may be of moment with regard to its astonishing capability of producing an enormous number of flowers consecutively. In fact, if the flowers are allowed to senesce, the stigmata being excised so that no seeds are developed, an individual plant produces on an average over 250 flowers as compared to only 174 if the flowers are removed immediately after anthesis. Concomitant with the degradation of wall polysaccharides various glycosidases show a marked increase in activity (Fig. 42; WIEMKEN-GEHRIG *et al.* 1974). The differential behaviour of tissues in the senescing corolla suggests that these enzymes may lyse the cell walls predominantly in the mesophyll. Another example of extensive degradation of hemicelluloses is shown in senescing tobacco leaf discs (MATILE 1974). The 14-fold increase of β-1,3-glucan hydrolase activity in senescing leaves of *Nicotiana glutinosa* observed by MOORE and STONE (1972) can perhaps be associated with hemicellulose degradation.

SACHER (1973) reviewed the more recent work which demonstrates the significance of cell wall lysis in the softening of fruits. This specific aspect of senescence has received special attention because it is important in food technology. Obviously, the softening of fruit tissues is caused by their disintegration and it is clear that this is brought about by the action of

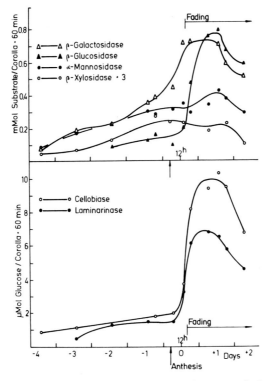

Fig. 42. Activities of glycosidases in the course of development of the *Ipomoea tricolor* corolla. After WIEMKEN-GEHRIG *et al.* (1974).

polygalacturonase, pectinesterase and even cellulase. In tomatoes all of these hydrolases, but particularly polygalacturonase are considerably more plentiful in ripening fruits as compared to green ones (HALL 1963, HOBSEN 1964, 1963, BESFORD and HOBSEN 1972). Similarly the softening of *Avocado* fruits is associated with a marked enhancement of polygalacturonase (ZAUBERMAN and SCHIFFMANN-NADEL 1972). Decreases of hemicellulosic polysaccharides have also been detected in ripening apples (KNEE 1973).

Abscission zones are specialized regions of petioles and pedicels in which separation of cells occurs in an advanced stage of senescence of leaves and fruits (Fig. 43). The examination of the fine structure of this tissue has shown that the separation of cells is initiated by the appearence of lysed areas in the middle-lamella; this phenomenon suggests the involvement of pectinolytic enzymes. In addition to these changes in the middle lamella which eventually lead to the breakdown of this wall layer (Plate 40) disintegration of the primary wall has also been observed (BORNMAN 1967, VALDOVINOS and JENSEN 1968). This circumstance points to the possible action of other cell wall lytic enzymes than pectinases in abscission.

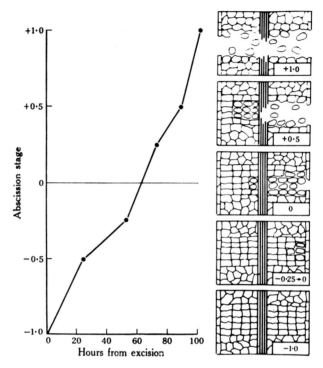

Fig. 43. Time course of cell separation in the abscission zone of *Coleus petioles.* DAVENPORT and MARINOS (1971) followed the process continuously by culturing slices of abscission zones. The sequence of abscission was divided into the arbitrary stages — 1.0 to + 1.0; the curve indicates scores of microscopically observed stages as a function of time.

It is somewhat surprising that most of the work on polysaccharidases as related to abscission has been carried out on cellulolytic enzymes rather than on pectinases. The reason for this is simply that pectinase activities assayed in extracts from abscission layers were either faint or not detectable at all (see RASMUSSEN 1973). However, RIOV (1974) recently succeeded in demonstrating the involvement of an exo-polygalacturonase in the separation of cells in citrus leaf explants (Fig. 44). HORTON and OSBORNE (1967), ABELES (1969), and SHELDRAKE (1970) convincingly demonstrated that cellulase activity increases in the separation zones of abscising *Phaseolus, Gossypium* and *Coleus* petioles. Similar changes in cellulase activity (external and internal), which were restricted to the separation zones, occurred in orange fruits induced to abscise (RASMUSSEN 1973). Abscission can be induced by the deblading of leaves; LEWIS and VARNER (1970) used debladed seedlings of *Phaseolus vulgaris* to investigate a possible correlation between cellulase activity in the pulvinus (the upper abscission zone of bean leaves) and the pulling force necessary to separate the abscission zone. The graph in Fig. 45 clearly demonstrates that less pulling force is required as cellulase activity begins to increase. The density labelling technique on explants of bean

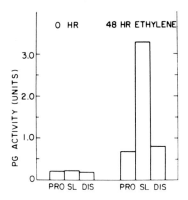

Fig. 44. Localization of polygalacturonase in abscission zone of *Citrus sinensis* leaf explants. The activity at zero time was determined on freshly excised explants (2 cm). The activity of ethylene-treated explants was determined after 48 hours incubation in 10 ppm ethylene. Each section represents 2 mm of tissue. *PRO* = the section proximal to the separation layer; *SL* = separation layer; *DIS* = the section distal to the separation layer. From RIOV (1974).

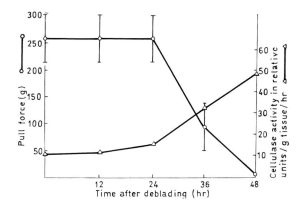

Fig. 45. Comparison of the changes in pull force (the force at which the abscission zone separated) and changes in cellulase activity with time after deblading primary leaves of *Phaseolus vulgaris* seedlings. From LEWIS and VARNER (1970).

abscission layers supplied with D_2O showed an increase in the buoyant density of cellulase in cesium chloride gradients (LEWIS and VARNER 1970). This finding indicates that cellulase is synthesized *de novo* upon induction of abscission. It seems that cell wall lysis in abscission layers requires a most precise spatial release of cellulase and perhaps other hydrolases. Little is known about the cytological aspect of this. Electron micrographs presented by VALDOVINOS et al. (1972) suggest the mediation of the rough ER and of the Golgi apparatus in vesicular secretion (Plate 40).

3.3.3. Cell Wall Lysis in Differentiating Cells

Anastomosis of cells in xylem or in compound laticifers comprises the partial dissolution of existing cell walls. In water-conducting vessels the removal or perforation of end-walls is a prerequisite for the free movement of water. Indeed the end-walls of hydroids in the moss *Dendroligotrichum* consist of the cellulose residues of a primary wall and it is thought that this is a result of specific hydrolysis occurring in the course of the differentiation of the hydroids (SCHEIRER 1973). The unhydrolyzed lateral-walls of hydroids seem to be protected from lysis through the action of polysaccharidases by incrusting lignin (Plate 38 A). Similar hydrolyzed walls have been observed in the primary xylem of higher plants (O'BRIEN 1970, O'BRIEN and THIMANN 1967, CZANINSKI 1972, BAL and PAYNE 1972). The functional significance of hydrolysed walls in pit membranes is evident, but it is surprising that the primary side walls of tracheary elements are also hydrolysed upon differentiation (O'BRIEN 1970, CZANINSKI 1972). In the differentiated tracheid this wall shows a fibrillar texture; it is cellulosic in nature and lacks acid polysaccharides. The ring-shaped secondary wall reinforcements of these cells are lignified and the rings are anchored, as demonstrated by O'BRIEN (1970) in willow, to similar bands formed by the neighbouring vascular parenchyma cells. O'BRIEN (1970) offered a sensible explanation for these phenomena: the lateral walls of tracheids must be stretchable so that these eventually dead cells can be expanded passively as the organ grows.

Hydrolysis of tracheary walls is most probably not organized by the living cell. It seems rather that the pattern of wall hydrolysis is determined by the lignification of those wall regions which are not to be attacked by the cell-wall lytic enzymes that are eventually released through the autolysis of the tracheary protoplast. Since these hydrolases would also attack the walls of neighbouring living xylem cells, these are protected by a special wall layer (O'BRIEN 1970). An account of cytological phenomena associated with these differentiations was presented by CZANINSKI (1968). It is not known whether specific structures are involved in the formation of the hydrolases which are destined to bring about cell wall hydrolysis after autolysis of the tracheary cytoplasm. A noticeable proliferation of rough ER-membranes appears to be associated with cell wall breakdown in cortical and provascular tissues of quiescent root meristems of *Allium cepa* (BAL and PAYNE 1972). Although cytochemical investigations suggest that cellulose is not hydrolyzed in the differentiated tracheids, SHELDRAKE's (1970) report on the occurrence of high cellulase activities in *Acer pseudoplatanus* xylem (and phloem) indicates that differentiation in these tissues may involve degradation of cellulose. Likewise, the occurrence of a highly active cellulase in *Hevea* latex suggests that this enzyme may play a role in the removal of cell wall material during the differentiation of the articulated laticifers of this plant (SHELDRAKE and MOIR 1970).

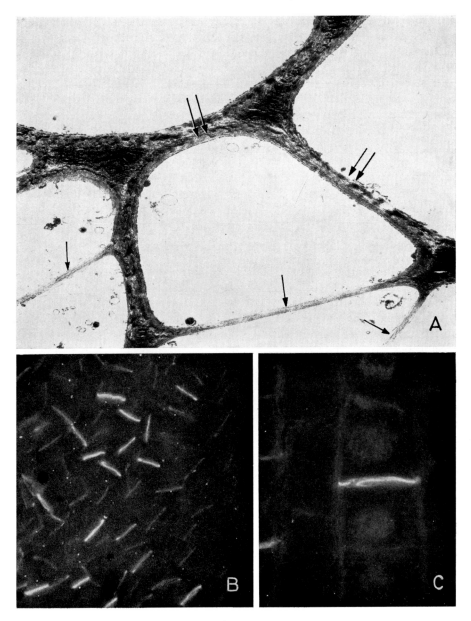

Plate 38.

A: Hydrolyzed walls in the water-conducting cells of *Dendroligotrichum* (Bryophyta). Electron micrograph of hydroids showing delicate hydrolyzed end-walls (arrows) in comparison to the "protected" lateral walls (double arrows). ×2,600. Courtesy of D. C. SCHEIRER. *B, C:* Cytochemical localization of callose: fluorescence upon treatment with aniline blue. *B:* Selaginella krausiana, basal portion of a developing leaf. Note fluorescent callose in the newly formed cell plates. ×750. *C:* Root meristem of *Hyacinthus orientalis* with telophase cells separated by a callosic cell plate. ×750. Courtesy of L. WATERKEYN.

This section should perhaps also contain a brief remark on a specific poly-saccharide which seems to play a particularly dynamic role in cell develop-ment: callose. It has already been mentioned in connection with the plugging of sieve tubes in *Vitis*. It is in fact widely distributed and serves as a plugging wall material not only in the specialized plasmodesmata of sieve pores but also in plasmodesmata generally. Its dynamism is documented by cyto-chemical studies of WATERKEYN (1967); what he calls "stades callosiques" are temporary depositions of this β-1,3-glucan, for instance in the cell plate after mitotic divisions in meristems (Plate 38 *B* and *C*). ABELES and FORRENCE (1970) showed that the activity of an endo β-1,3-glucanase present in explants of bean petioles in the abscission zone is inversely correlated with the amount of callose present in sieve tubes, suggesting that this enzyme plays a role in the removal of sieve pore plugs.

3.4. Cell Wall Degradation in Parasitism

A most fascinating complex of cell wall lytic processes and involvement of lytic cell compartments, both of the pathogenic fungus and of the host cells that it invades, will be discussed here.

It is evident that a parasitic fungus which utilizes cellular substances of the host plant to cover its nutritional requirements has to force the host cells open in one way or another. This can be accomplished in various ways other than through the action of hydrolase, for instance by the release of toxins which affect the membranes of the host cells or by the use of sophisticated mechanisms that penetrate the cell walls of the host mechanically. In many pathogenic fungi the lysis of the cell walls of the host appears, however, to be brought about by secreted hydrolases. Indeed, the elucidation of ultra-structural changes following fungal invasion has yielded consistent evidence that the cell walls of the host are perforated at the sites of penetration of the hyphae (Plate 37 *C; e.g.,* HESS 1969, LITTLEFIELD and BRACKER 1970, POLITS and WHEELER 1973), extensive wall degradation may even be caused by certain pathogens (CALONGE *et al.* 1969). Moreover, ultrastructural features of the germ tubes of *Botrytis fabae* penetrating into *Vicia faba* mesophyll cells (Fig. 46) suggest that lomasome-like bodies may be involved in the local extracellular release of cell wall lytic enzymes (ABU-ZINADA *et al.* 1973). However, similar structures, evidently produced by the host in response to the presence of *Puccinia* have also been observed in wheat leaves (EHRLICH *et al.* 1968).

Such ultrastructural phenomena as lomasome formation have not yet been correlated with biochemical events taking place upon fungal invasion in the case of either of the interacting organisms. However, abundant data on the involvement of the external lytic compartment of the parasite is available.

WOOD (1967) compiled the numerous cases of plant pathogenic fungi that are able to secrete various cell wall lytic enzymes which specifically hydrolyze components of plant cell walls: cellulases, pectic enzymes, hemicellulases and proteinase. The effect of pectic enzymes is particularly obvious since they cause maceration of the host tissues (see BATEMAN and MILLAR 1966). Tissue degradation through the action of macerating enzymes such as the "phytolysin" isolated by KERN and NAEF-ROTH (1971) in conjunction with other cell wall lytic enzymes may result in the loss of the rigidity of cell walls to such an extent that they no longer resist the turgor pressure. Indeed, purified pectic enzymes produced by *Erwinia chrysanthemi* are able to initiate the killing of host cells (GARIBALDI and BATEMAN 1971). As a con-

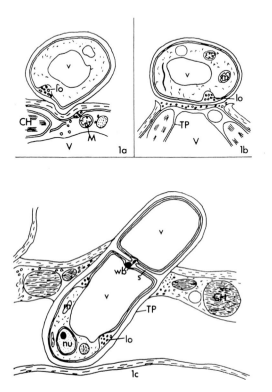

Fig. 46. Diagram showing the ultrastructural features of *Botrytis fabae* germ tubes penetrating into *Vicia faba* mesophyll cells. Lomasomes (*lo*) are possibly involved in the lysis of the host cell wall. From ABU-ZINADA *et al.* (1973).

sequence of the weakening of the walls the cells burst and lesions are produced in the infected plant organ (*e.g.*, MUSE *et al.* 1972, HAGAR and McINTYRE 1972). For instance, it has been demonstrated that the capability of *Scleroticum bataticola* to produce polygalacturonate-transeliminase and cellulase whilst growing in host tissue is closely correlated with the virulence of the strains tested (CHAN and SACKSTON 1972). However, in addition to these enzymes, other hydrolases are important in the penetration of this fungus into sunflower stems—and this brings us back to the interesting features of the infection process.

According to work done by ALBERSHEIM and his group, "polysaccharide degrading enzymes are unable to attack plant cell walls without prior action by a wall-modifying enzyme" (KARR and ALBERSHEIM 1970). This enzyme was first purified from a commercial preparation of pectic enzymes from *Aspergillus niger;* it has a limited ability to degrade polygalacturonic acid, however, it modifies the structure of isolated *Phaseolus* hypocotyl walls sufficiently to make other polysaccharides accessible to the corresponding hydrolases. An enzyme of this type has also been isolated from the culture filtrate of *Colletotrichum lindemuthianum* grown on citrus pectin as a source of carbon (ENGLISH *et al.* 1972). The decisive fact with regard to the role of this polygalacturonase is, however, that it is the first to be secreted, when the pathogenic fungus is grown on cell walls isolated from a host tissue susceptible

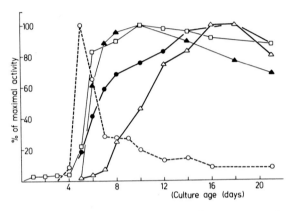

Fig. 47. Hydrolase secretion in *Colletotrichum lindemuthianum* grown in a medium containing cell walls isolated from Red Kidney bean hypocotyls as the sole carbon source. Polygalacturonase (O———O), β-xylosidase (□———□), cellulase (▲———▲), β-glucosidase (●———●), and α-galactosidase (△———△) are secreted sequentially. Slightly simplified from ENGLISH *et al.* (1971).

to infection (ENGLISH *et al.* 1971). Moreover, it appears from the levels of various polysaccharidases that appear in the culture medium at different times (Fig. 47) that a sophisticated program of enzyme release is established in *C. lindemuthianum* guaranteeing the eventual lysis of walls and certainly the killing of host cells. A similarly programmed secretion of polysaccharide-degrading enzymes has been detected in *Fusarium oxysporum* grown on cell walls isolated from tomato stems (JONES *et al.* 1972). Most probably not the "wall modifying enzyme" alone is responsible for the successful attack, but the total of the organized army of enzymes produced by the pathogen. In the case of one hydrolase, α-galactosidase, which is secreted last, a positive correlation of its production in various strains of the pathogen with their virulence has been noticed (ENGLISH and ALBERSHEIM 1969). In addition, there is a significant response in *C. lindemuthianum* to the origin of cell walls employed for *in vitro* cultures. Leaves, hypocotyls and internodes of bean plants are infected and cell walls isolated from these organs induce the fungus to secrete α-galactosidase; in contrast, roots are resistant to infection and their cell walls fail to induce high activities of this enzyme (ENGLISH *et al.* 1971).

A study of KEEGSTRA et al. (1972) shows that in C. lindemuthianum several glycosidases are selectively induced in the presence of the corresponding sources of carbon, e.g., α-galactosidase in the presence of galactose; since small amounts of such inducing sugars seem to be liberated in isolated cell walls upon treatment with the "wall modifying" endopolygalacturonase (ENGLISH et al. 1972) these may trigger the secretion of additional polysaccharidases. It is noteworthy that Colletotrichum releases proteolytic activity when cultured either on cell walls alone (PLADYS et ESQUERRÉ-TUGAYÉ 1973) or on collagen as a source of nitrogen (RIES and ALBERSHEIM 1973). May be this hydrolase degrades protein in primary walls or the cytoplasmic protein which becomes accessible after the cells of the host are ruptured.

After having considered how pathogenic fungi successfully attack and dissolve the cell-walls of the host the question of how host cells may also be resistant to infection can be raised. A possible explanation of resistance has again been worked out in ALBERSHEIM's laboratory. FISHER et al. (1973) were able to isolate a protein from Phaseolus vulgaris which efficiently inhibits polygalacturonase secreted by the pathogenic fungus Colletotrichum lindemutianum. Although nothing is known about the cellular location of this inhibitor protein (which can be expected to be present in cell walls) it is tempting to assume that it blocks the fungal polygalacturonase and thereby makes infection impossible. However, the hypocotyls of three varieties of red kidney bean differing in susceptibility to infection contained the inhibitor protein in comparable amounts (ANDERSON and ALBERSHEIM 1972). There was also no correlation between the pathogenicity of strains of C. lindemutianum and their ability to produce endopolygalacturonase. Since the fungus produces proteinase it can be hypothesized that it is able to overcome an infection barrier of extracellular inhibitor proteins of the host. Another possible explanation of resistance to infection is that the host turns the tables by secreting polysaccharidases capable of lysing the invading fungal hyphae. Indeed, ABELES et al. (1970) isolated an endo-β-1,3-glucanase and a chitinase from bean leaves. The activity of these enzymes increases dramatically when the leaves are exposed to ethylene the production of which is known to be associated with fungal infections (WILLIAMSON 1950). Hence, the host plant seems in fact to produce enzymes capable of lysing fungal cell walls. To make the picture of attack and defense in host-pathogen interaction even more colourful ALBERSHEIM and VALENT (1974) recently discovered specific inhibitor proteins secreted by Colletotrichum lindemutianum which efficiently inactivate the polysaccharidases produced by the host plant to attack the pathogen.

In 1969 ALBERSHEIM et al. proposed a hypothesis of pathogenesis stating that "it is an interaction between the pathogen and the carbohydrates of the host which determines the pathogen's ability to produce enzymes capable of degrading the host's cell walls". Not only did ALBERSHEIM bring to light a number of biochemical facts which support this view; his recent findings suggest that whether or not a host plant tissue is infected depends on the balance of power inherent in the hydrolases and corresponding inhibitors produced by either of the interacting organisms.

Not all the parasitic fungi harm their hosts by killing cells. There are numerous infections in which the fungus develops a haustorium inside the host cell (review by BUSHNELL 1972). It looks as if this were the result of a balance between cell wall lysis caused by the fungus and cell wall repair by the host cells (Fig. 48). Such a form of parasitism appears to be more effective than infection associated with the killing of host cells. The same phenomenon can be observed in *Cuscuta*, a parasitic higher plant which is able to penetrate the

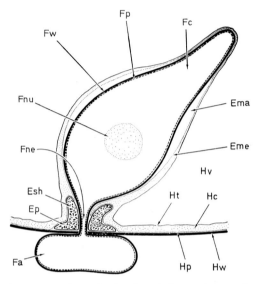

Fig. 48. Diagrammatic cross-section of the fungal haustorium and associated structures of a higher plant. *F* components of the fungus; *H* components of the host cell. Infection involves the lysis of the host cell wall (*Hw*). Papillae and sheaths (*Ep, Esh*) are thought to represent attempts of the host to wall out the parasite. From BUSHNELL (1972).

host cells with hypha-like organs (see DÖRR 1969). The development of the haustorium of *Orobanche* seems also to involve cell wall degradation in the host-tissue (DÖRR and KOLLMANN 1974). Indeed, the intensity of β-glucosidase activity demonstrated cytochemically in host-parasite unions of a tropical semiparasitic *Loranthaceae* suggests the involvement of cell wall lysis in the development of mistletoes in branches of their host-trees (ONOFEGHARA 1973).

3.5. Lysosomal Involvement in Plant Pathology

Infection of plant tissues by pathogenic fungi and other microbes causes a variety of changes in the host cells. Lytic phenomena concerning the cell walls have already been discussed (see section 3.4.). Changes suggesting an involvement of the internal space of the lytic compartment have also been reported.

One of the most conspicuous responses to inoculation of plant tissues with pathogenic microbes is the increase in RNase activity (*e.g.*, ROHRINGER *et al.*

1961, REDDI 1966, BRAUN and WOOD 1966, JOHNSON et al. 1968, CHAKRA-
VORTY and SHAW 1971). This reminds one of similar enhancement of RNase
levels in senescent tissues. However, the decline of RNA content which
occurs in the course of senescence has not been found in infected tissues. It
would, therefore, not be justified to regard infection as an induced premature
ageing, yet it is possible that the products of RNA-breakdown in the host
pass to the parasite.

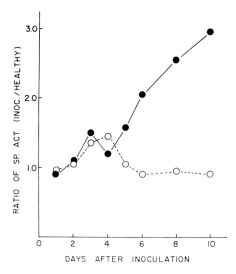

Fig. 49. RNase activity of resistant and susceptible flax cotyledons at different times after
inoculation with flax rust. O———O resistant flax (*cult*. Bombay); ●———● susceptible
flax (*var*. Bison). From SCRUBB et al. (1972).

According to SCRUBB et al. (1972) a small but significant increase in RNase
activity occurs in a resistant cultivar as well as in a susceptible one of flax
when cotyledons are inoculated with flax rust; however, a second much larger
increase in RNase activity could be detected only in the susceptible plants
(Fig. 49). A similar enhancement of RNase activity has been observed in rust
infected wheat leaves (CHAKRAVORTY et al. 1974). It is tempting to assume
that this response is caused by tissue damage due to fungal invasion in the
susceptible tissue. In fact, mechanical damage induces a similar enhancement
of RNase activity in flax cotyledons; however, SCRUBB et al. (1972) observed
qualitative differences in comparison with the effect of infection. In addition,
mechanical injury induces increased activity of a variety of hydrolases
(RNase, DNase, phosphodiesterase, phosphatases) whereas infection apparently has a specific effect on RNase only. Using an immunochemical test for
the quantification of RNase-protein (instead of RNase activity which is
normally assayed) PITT (1971) showed that the increase in RNase activity
following damage to potato tuber tissue is likely to be due to activation of
preexisting enzyme protein rather than to the formation of new. This appears
to be true also for the enhancement of RNase activity associated with fungal

9*

infection (PITT and GALPIN 1973). Hence, there are, at least in the case of potato, similarities between the changes caused by mechanical damage and disease. Recently, PITT (1974) showed that both activation of preexisting RNase and *de novo* synthesis of this enzyme occur in mechanically damaged potato leaves.

Histochemical tests performed by PITT and COOMBES 1968, 1969) showed that rotting of tuber tissue of *Solanum tuberosum* induced by *Phytophtora erythroseptica* results in the disrupture of lysosome-like organelles. The phase-dense granules react positively in a test for acid phosphatase and unspecific

Table 5. *Acid Phosphomonoesterase Activity in Particulate and Supernatant Fluid Fractions of Potato Callus Tissue Infected with Soft Rot Fungi* (after PITT and COOMBES 1969).

Treatment	% of total activity in	
	particulate fraction	supernatant fraction
Phytophtora erythroseptica		
uninfected	57.2	42.8
4-day infected	31.0	69.0
Phytophtora infestans		
uninfected	57.4	42.6
5½-day infected	22.5	77.5
Fusarium caeruleum		
uninfected	68.1	31.9
4½-day infected	28.9	71.1

esterase; since they can be stained with neutral red it is possible that they are a kind of small vacuole. During infection they swell and eventually burst (Plate 39). PITT (1973) succeeded in isolating membrane-bound structures from potato leaf homogenates. Infection of leaves by *Phytophtora infestans* induced a conspicuous transfer of a specific phosphatase isoenzyme to the soluble cell fraction; in homogenates from healthy leaves this isoenzyme was present in a latent form in vesicles. A similar solubilization of acid phosphatase has been demonstrated in *Phytophtora*-infected callus tissue of potato (Table 5; PITT and COOMBES 1969). This is obviously the biochemical correlate of the disrupture of lysosome-like granules which can be demonstrated histochemically in the healthy tissue. In my opinion the annihilation of hydrolase compartmentation upon the induction of soft-rot diseases is an unspecific response of the infected host tissue, which is caused by cell wall degradation. The osmotic lysis of cells and concomitant bursting of membrane-bound cell compartments such as vacuoles is likely to be ultimately caused by the release of cell wall degrading enzymes by the invading fungus.

Invasion of tuber cells of orchid by mycorrhiza eventually leads to the digestion of the endophytic hyphae. WILLIAMSON (1973) observed that during infection of host cells a conspicuous increase of the number of acid phosphatase positive granules occurs shortly before the hyphae lyse. Since phos-

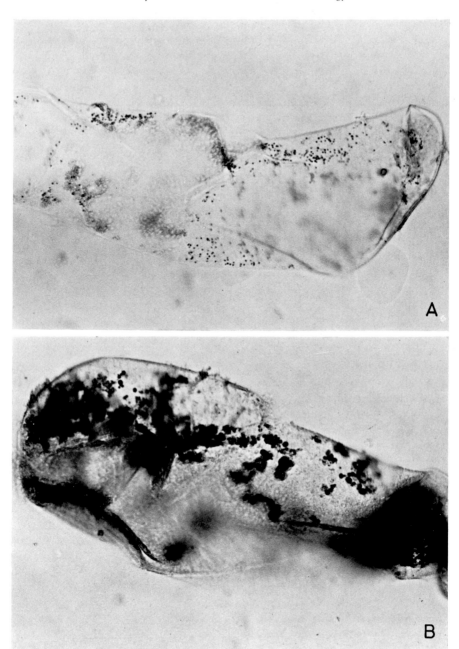

Plate 39. The disruption of lysosome-like particles of *Solanum tuberosum* cells during infection by *Phytophthora erythroseptica*.

A: Callus cells stained for acid phosphatase. Note many discrete acid phosphatase-positive particles. ×500. *B:* Callus cell 48 hours after infection with discs of *Phytophthora erythroseptica*. Note enlarged acid phosphatase-positive particles some of which have coalesced; also intense diffuse staining. ×500. Courtesy of D. PITT.

phatase is also present in granules in the fungal cytoplasm before lysis, it is uncertain whether the diffuse reaction seen in the lysed fungus is due to enzyme secreted by the host cell. A striking proliferation of rough-surfaced ER in the infected *Neottia* root (DÖRR and KOLLMANN 1969) supports the view that production of secretory enzymes takes place in the host cells. In conclusion, the experimental difficulties inherent in research on the problems of the interactions between two organisms will probably delay progress in the understanding of lysosomal involvement in plant pathology. A preview on this subject was recently published by WILSON (1973).

4. Control of Lysis

As lytic processes constitute an integral part of plant metabolism they must be rigorously controlled. In fact, the existence of control appears from a large number of studies in which the involvement of phytohormones and other factors in the regulation of cellular breakdown reactions is demonstrated. There is no doubt that in plant development the catabolic processes are as precisely controlled as are the anabolic. However, at the present time we are not yet in a position to recognize the totality of regulatory principles involved.

The most obvious control of lysis is given by the compartmentation of hydrolases. Uncontrolled lysis of intracellular constituents seems to occur only after the disrupture of the lytic compartment in the autolysing cell. The separation in the living cell of hydrolases and their potential substrates suggests that hydrolase activities assayed in homogenates do not necessarily reflect the rates of breakdown reactions. The investigation of control at the level of hydrolase activities will, therefore, shed light upon only one aspect of lysis. It is true, that the assessment of degradation rates will fill the gap. However, it will be impossible to elucidate the cytological aspects of control as long as only biochemical parameters of lysis are considered. An important consequence of hydrolase compartmentation is that control of lysis also comprehends the control of membrane processes. Autophagic activity as well as hydrolase secretion, which are responsible for the contact between digestive enzymes and cellular substances to be digested, involve growth and differentiation of membranes. At the present time nothing but isolated facts towards the understanding of the control of lysis from biochemical and, to a minor extent, cytological viewpoints are available. However, in most cases it is not yet possible to put these facts together to make up a comprehensive pattern of how lysis is controlled in plant cells.

4.1. Seed Germination

The availability of sufficient moisture is the most obvious condition for the initiation of seed germination. Admittedly, this is a truism, yet water has an immediate relationship with the hydrolysis of stored macromolecules. It is a reagent, not only the medium in which seeds are soaked. As in the quiescent seeds proteinases capable of initiating the mobilization of storage proteins are already present, mobilization begins immediately upon the penetration of water into the reserve tissues. Hence, water controls the initial lysis in seeds.

Free amino acids that accumulate upon proteolysis seem to regulate the proteinase level in storage organs of seeds for instance in cotyledons of germinating peas (YOMO and VARNER 1973). Although this phenomenon does not reflect a direct inhibition of proteinase activity in the presence of high levels of amino acids, it nevertheless demonstrates that, in addition to the control of protein mobilization by water, control is also excercised by the developing tissues of the embryo. Evidently, the lytic processes in storage tissues are controlled by the embryonic organs which utilize the amino acids and other products of lysis. It appears from studies with detached pea cotyledons that the total protein practically does not decline upon inbibition whereas in attached cotyledons it gradually decreases as germination proceeds (CHIN et al. 1972). In detached cotyledons of *Cucurbita maxima* the amount of protein hydrolysis is comparable with hydrolysis in the attached organs if excised embryonic axes are added to the incubation medium (WILEY and ASHTON 1967). Lysis in the cotyledons appears therefore to depend upon a stimulus which is sent out by the axis; it is under *hormonal control.* The mobilization of protein in cotyledons corresponds with protein degradation in other senescing leaves in that it is also regulated by hormones produced in the developing tissues of the plant.

A classical object for the study of hormonal control of mobilization in germinating seeds is the barley caryopsis. As early as 1890, HABERLANDT reported the release of starch liquefying enzymes from the aleurone layer in response to substances produced by the germinating embryo. In half seeds without embryos or in isolated aleurone layers this effect of the embryo can be replaced by gibberellic acid. In addition to amylase, a variety of other hydrolases such as proteinase, ribonuclease, phosphatase, β-glucanase, and pentosanase are secreted in response to this phytohormone (review by YOMO and VARNER 1971).

Amylase, proteinase, glucanase and RNase are synthesized *de novo* in barley aleurone layers treated with gibberellic acid (CHRISPEELS and VARNER 1967, JACOBSEN and VARNER 1967, BENNETT and CHRISPEELS 1972). An ingenious technique used for demonstrating *de novo* synthesis is density labelling of protein. In the presence of isotopically labelled water ($H_2^{18}O$) the hydrolysis of storage proteins yields labelled amino acids which are subsequently used in the synthesis of these hydrolases; the increased buoyant density in CsCl gradients of labelled hydrolases demonstrates unequivocally that these enzymes were synthesized *de novo* from amino acids produced upon the hydrolysis of storage proteins (FILNER and VARNER 1967). Since inhibitors of protein synthesis, which are known to act at the level of DNA transcription, impair the formation of amylase, the dramatic effects of gibberellic acid were interpreted in terms of differential gene activation. However, long-lived messenger RNA coding for α-amylase seems to be present in aleurone cells before gibberellic acid acts upon its target tissue (CARLSON 1972). Moreover, the first hydrolase to be secreted in hormone treated aleurone layers, β-glucanase, is synthesized in the absence of gibberellic acid. This enzyme accumulates in the cells if half seeds without embryos are imbibed with water. Secretion of β-glucanase requires the exposure of the aleurone layers to gibberellic acid (JONES 1971). This phenomenon suggests that the hormone has an effect on membrane processes involved in hydrolase secretion rather than exclusively on the synthesis of these enzymes. Indeed, JOHNSON and KENDE (1971) detected a marked enhancement, dependent on gibberellic acid, of certain enzymes that participate in the synthesis of lecithin. This enhancement clearly precedes the secretion of β-glucanase. It can be concluded from these observations that the effect of gibberellic acid on the barley aleurone cells is a multiple one. The early effect on the synthesis of phospholipids (KOEHLER and VARNER 1973) points to the formation of membranes which are subsequently involved in the synthesis and subsequent secretion of hydrolases. Morphological observations made by JONES (1969 b) on the development of rough surfaced ER as well as corresponding biochemical results (EVINS and VARNER 1971, 1972) during the lag phase of amylase secretion support this view. Hence, the hormonal control of hydrolase secretion in the barley alleurone cells appears as a complex of coordinated processes which are induced in response to the stimulus sent out by the embryo. It should be mentioned that this control is possibly not effected by gibberellic acid alone; an interaction between this hormone, abscisic acid, and ethylene in the control of amylase synthesis in barley aleurone layers was recently reported by JACOBSEN (1973).

The hormonal control of lytic processes in the barley endosperm is characterized by comparatively slow responses to induction by gibberellic acid. A period of about 10 hours elapses before amylase and proteinase appear in the bathing solution of induced aleurone layers (Fig. 51). It also seems that quick changes in the rate of hydrolase secretion cannot be evoked by changes in the flow of gibberellic acid from the embryo to the target tissue: the synthesis and release of hydrolases in isolated aleurone layers is roughly proportional to the logarithm of gibberellic acid concentration which means that changes in hormone production in the embryo have little influence on the secretory activity of the aleurone cells. In fact, JONES and ARMSTRONG (1971) observed that in intact barley seeds the level of gibberellic acid continues to increase after the level of α-amylase has reached a maximum 3 to 4 days after germination. It appears then, that the hormonal induction has almost the character of an all-or-nothing-effect and additional mechanisms may be involved in a more precise regulation of reserve mobilization. Such a mechanism has, in fact, been elucidated by JONES and ARMSTRONG (1971) who produced evidence for an osmotic regulation of hydrolase production in germinating barley seeds. Upon the liquefaction of starch in the starchy endosperm large quantitites of sugars, mainly glucose and maltose, accumulate. If isolated and induced aleurone layers are treated with solutions of sugar or polyethylene glycol at increasing osmolarities the amylase production is progressively reduced. Incidently, the osmotic treatment of gibberellic acid induced aleurone layers causes a progressive reduction of the binding of ribosomes to the ER (ARMSTRONG and JONES 1973). The concentration of

osmotically active solutes and enzyme release were found to be roughly proportional. This means that the doubling of the concentration of the solute
results in an immediate reduction of amylase production by up to 50%,
whereas a corresponding change of gibberellic acid level causes only a slow
change of ultimately 5%. Thus, osmotic regulation of hydrolase production
in conjunction with hormonal induction appears to be a fine system of complex
control in which, of course, the absorption and processing of endosperm solutes

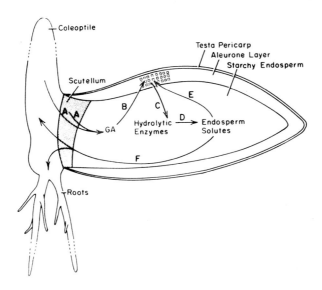

Fig. 50. Diagrammatic representation of the relationship between gibberellic acid production,
α-amylase production and solute accumulation in germinating barley seeds. Gibberellic acid
produced by the coleoptile and scutellum (A) migrates into the aleurone layer (B) where
hydrolytic enzyme synthesis and release is induced (C). These enzymes serve to hydrolyze
the reserves of the endosperm (D), producing solutes which inhibit further hydrolase
production (E), and function in nourishing the growing embryo (F). From Jones and Armstrong (1971).

in the scutellum is a further important link (Fig. 50). Evidence was given by
Radley (1969) which supports the view that sugar metabolism in the scutellum
in turn influences the gibberellin synthesis in the embryonic axis.

It appears from Fig. 51 that ribonuclease is the last hydrolase to be released
from induced barley aleurone layers. In the course of the lag phase which
elapses until its activity appears in the bathing solution, the aleurone cells
become extensively vacuolated (Jones and Price 1970). According to Gibson
and Paleg (1972) RNase is an apparently soluble enzyme, in contrast to
amylase and proteinase which are associated with sedimentatable structures.
It may be speculated that RNase actually is localized in the vacuoles of
aleurone cells, its release occurring in connection with autolysis which presumably takes place towards the end of endosperm mobilization. The release
of RNase is reduced in the presence of high solute concentrations, indicating

that under these conditions the vacuoles may be osmotically stabilized. As the walls of aleurone cells are extensively degraded the vacuoles may burst and vacuolar RNase be released in the absence of osmotic stabilization.

VARNER and MENSE (1972) were able to distinguish experimentally between *secretion*—the outward movement of amylase through the plasmalemma— and *release*—the movement through the cell walls of aleurone layers into the starchy endosperm. The release of α-amylase was found to be markedly dependent upon the presence of Ca^{++}, Mg^{++} and K^+. The activation of enzymes such as phytase (EASTWOOD et al. 1969) in induced aleurone layers

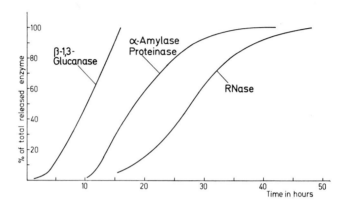

Fig. 51. The time courses of release of hydrolases from gibberellic acid treated barley aleurone layers. Computed from TAIZ and JONES (1970), JONES and PRICE (1970), and JACOBSEN and VARNER (1967).

has, therefore, an important aspect not only in the supply of ions to the embryo but also in overcoming the possible ionic bounds holding the secreted hydrolases to the aleurone cell wall.

To end up this rather extensive account on control mechanisms in the germinating barley grain it should be mentioned that the involvement of hormones such as gibberellic acid, cytokinins, abscisic acid and ethylene in the mobilization of storage products in seeds of dicotyledons is documented by a large number of reports which shall not be cited here. As a rule gibberellic acid or cytokinin can replace the promoting stimuli produced by embryonic tissues of germinating seeds, whereas abscisic acid and ethylene appear to have antagonistic effects. In *Sinapis alba* the mobilization of the cotyledonary storage protein is controlled by the phytochrome system (HÄCKER 1967). This is concluded from the promoting effect of red light which can be reversed by far-red irradiation. In apple embryos grown in the dark the activity of alkaline lipase is stimulated by red irradiation (but is not increased by far-red irradiation) as well as by treatment with gibberellins (SMOLÉNSKA and LEWAK 1974). Therefore, the phytochrome system may control light dependent degradative processes in germinating seeds indirectly by influencing the levels of gibberellins.

4.2. Growth and Differentiation

The significance of cell wall loosening through the action of hydrolases for extension growth has been covered in section 3.3.3. As cell elongation in internodes is under hormonal control which is exercised by the apex, it is not surprising that effects of auxin on the synthesis of cell wall degrading hydrolases have been detected. Cellulase activity in the epicotyls of decapitated pea seedlings was found to increase concomitant with growth which was induced upon the application of indoleacetic acid. Inhibitory effects of actinomycin D and cycloheximide on the auxin-dependent production of cellulase, β-1,3-glucanase and pectic enzymes suggest that the hormonal regulation of cell wall plasticity is associated with protein synthesis (FAN and MACLACHLAN 1967, DATKO and MACLACHLAN 1968). DAVIES and MACLACHLAN (1968) reported that cellulase activity generated under the influence of auxin is localized in isolated cell walls and, notably, in a microsomal fraction prepared from pea epicotyls. Moreover, microsomes from auxin-treated tissue incorporate amino acids into protein at a much higher rate than do microsomes from untreated epicotyls, and even in an *in vitro* system of protein synthesis the increase in cellulase activity is noticeable (DAVIES and MACLACHLAN 1969). Whether or not these results on the hormonal control of cellulase production in growing tissues will be confirmed, an important problem concerning the regulation of cell wall lytic enzymes after their secretion into the wall remains. Cell walls contain a variety of glycosidases and it would seem that the cell must have a means of controlling the hydrolysis of cell wall polysaccharides according to the requirements of growth. It was shown by RAYLE (1973) that auxin induces the secretion of hydrogen ions in *Avena* coleoptiles.

The exposure of this tissue to acid solutions causes a significant enhancement of cell wall loosening and elongation of cells in the absence of auxin (RAYLE and CLELAND 1970). A recent report by JOHNSON *et al.* (1974) suggests that this effect is due to the activation of glycosidases associated with the cell wall which are characterized by acid pH optima around 5; in segments of *Avena* coleoptiles the auxin induced acidification of the mural space provokes a significant activation of β-galactosidase as assessed *in vivo*. Hence, the regulation of lysis is not necessarily dependent only on the synthesis of hydrolases, it can also be achieved through the control of pH in the lytic compartment. EVANS (1974) has, however, recently ruled out the possibility of a direct action of exo-glycosidases in auxin- or acid-promoted growth; specific inhibitors of α-galactosidase and β-glucosidase do not affect growth. This result suggests that endo-glycosidases may be chiefly involved in the wall loosening, the exo-glycosidases merely being responsible for the breakdown of fragments of polysaccharides produced by the action of these enzymes.

A hormonal control of enzymatic wall softening in the aquatic fungus *Achlya* was elucidated by THOMAS and MULLINS (1967). Sex hormone A which induces the branching of hyphae in the male sex organs also induces the enhancement of cellulase activity. Cellulose is one of the principal cell wall polysaccharides in *Achlya*, a genus of the *Saprolegniales*. The peak of the hormonally induced *de novo* synthesis of cellulase corresponds in time with the appearance of branches. Hence, there is little doubt that branching involves a controlled cell wall lytic process. Vegetative branching is also accompanied by a rise in cellulase activity; in contrast to antherical branching it is, however, induced by exogenous amino acids.

Extension growth is associated with enlargement of the vacuoles which eventually coalesce into the central vacuole of the fully expanded cell. The example of laticifers of *Euphorbia characias* demonstrates that vacuolation may be, in turn, associated with an extensive autophagic breakdown of cytoplasmic material (MARTY 1970). This and similar phenomena observed in other tissues point to cell differentiation which accompanies cell expansion. However, very little is known about the biochemical correlate of autophagy in expanding cells. It is evident that it must be under the control of an agent which guarantees the organized development of cells within a tissue. Experimental approaches to this problem are difficult, since there is a marked heterogeneity of behaviour of individual cells according to their prospective specific functions. A few reports deal with hormonal control of lytic activities which may be involved in the autolytic breakdown of cellular constituents in expanding tissues. During the growth of bean leaves the level of alkaline proteinase is raised (RACUSEN and FOOTE 1970). Growth promoting doses of the synthetic auxin 2,4-D cause an enhancement of proteinase activity in soybean internodes (FREIBURG and CLARK 1955). The control of RNase activity influenced by phytochrome in etiolated hypocotyls reported by ACTON (1972) is another piece of evidence for the control of lytic activity in growing tissues. The involvement of membranes in autophagic processes indicates that the regulation must also comprise the organized transfer of cytoplasmic material into the lytic compartment. Turnover of protein as the resultant of protein synthesis and (autophagic?) degradation are in fact under hormonal control as shown by TREWAVAS (1972 b) in growing *Lemna minor*. Depending on whether or not the culture medium contains sucrose, the cytokinin benzyladenine alters the rate constant of protein synthesis only, or, on a mineral medium, the rate constant of degradation only. On the other hand abscisic acid alters both the rate constants of synthesis and of protein degradation. According to TREWAVAS (1972 b) conditions which slow down growth, reduce the rates of protein synthesis and increase the rates of degradation. It would be very interesting to learn something about the possible morphological correlates of autophagy in this organism under conditions provoking different degradation rates.

4.3. Control of Senescence and Abscission

The rapid loss of chlorophyll, protein, RNA and other constituents which occurs in detached leaves can be reversed or prevented if adventitious roots are allowed to develop on the petiole. MOTHES (1960) and his co-workers discovered that the effect of juvenility factors which are produced by roots and transported into the leaves can be replaced by kinetin. The application of this compound to detached leaves is thought to replace the natural supply of cytokinins produced in the roots. Since the pionier work of MOTHES this retardation of leaf senescence, particularly the delayed breakdown of protein and RNA caused by cytokinins, has been demonstrated in various plant species.

KENDE (1971) thoroughly discussed the question of whether the effect of cytokinins concerns a stimulation of the synthesis of protein and RNA or an inhibition of breakdown processes. A delay in breakdown would be the result of either of these possible cytokinin effects. A clear-cut decision is prevented by the lack of exact assessments of turnover rates due to experimental diffi-culties. However, there is ample evidence to support the hypothesis that cytokinins delay senescence by inhibiting breakdown processes (see KENDE 1971). This hypothesis is indirectly supported by the fact that the levels of proteinase (e.g., ANDERSON and ROWAN 1966, BALZ 1966, BEEVERS 1968, ATKIN and SRIVASTAVA 1969, MARTIN and THIMANN 1972) and RNase (e.g., SRIVASTAVA and WARE 1965, BALZ 1966, SRIVASTAVA 1968, SODEK and WRIGHT 1969, WYEN et al. 1972, ARAD et al. 1973, see also DOVE 1973) are lower in detached leaves treated with cytokinin as compared to the untreated controls. The possible involvement of the lytic compartment in the hormonal regulation of breakdown processes in senescing tobacco leaves appears from the investigations of BALZ (1966). As shown in Fig. 52 the application of kinetin largely suppresses the increase in RNase and proteinase. In the untreated detached leaves these hydrolases increase first in a sedimentatable fraction of the homogenate and then in the soluble fraction, suggesting that some sort of primary lysosome is formed and then discharged into the vacuoles; although nothing is known about the cellular location of "soluble" RNase in tobacco leaves it may be speculated that it is localized in vacuoles. Hence, kinetin may not only suppress the synthesis of hydrolases but also the membrane processes involved in the supplementation of the lytic compart-ment with digestive enzymes.

It should be born in mind that hydrolase levels do not necessarily reflect the rates of degradation of cellular macromolecules. The concept of the lytic compartment implies that digestive enzymes determine rates of breakdown only in conjunction with autophagic activity, that is with the introduction of cytoplasmic material into the lytic compartment. Investigations of the morphological changes associated with senescence have mainly concerned the chloroplasts the degeneration of which is delayed in the presence of cytokinins. The effect of hormone treatments upon autophagic activities in mesophyll cells has not particularly been taken into consideration so far. However, in detached leaves treated with kinetin a delay of the structural changes occurring in senescing leaves has been observed (e.g., SHAW and MANOCHA 1965, MITTELHEUSER and STEVENINCK 1971).

Cytokinin is not the only factor responsible for the control of lysis in senescent leaves. Similar effects in preventing or delaying catabolic processes have been attributed to light, sugar supply, and auxin. According to SACHER (1969) auxin prevents the reduction of RNA levels in explanted bean endo-carps and causes a decrease in RNase. PILET (1970) reported a corresponding effect of auxin in excised rootlets of Lens culinaris, and an auxin mediated control of RNase and acid phosphatase in pieces of *Rhoeo* leaves appears from a study of DE LEO and SACHER (1970).

Promoting effects on senescence have been attributed to abscisic acid and, notably, to ethylene. Most spectacular among phytogerontological effects of

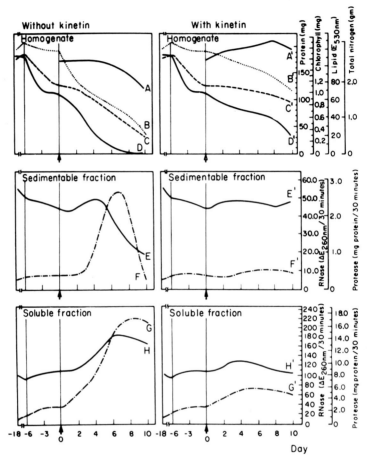

Fig. 52. The activitiy of hydrolases in kinetin treated and control halves of tobacco leaves. Mature leaves were detached from the plant on day 0 and one-half of each blade was sprayed with kinetin while the other half served as control. *A, A′* total nitrogen; *B, B′* lipids; *C, C′* protein; *D, D′* chlorophyll; *E, E′* sedimentable proteinase, *F, F′* sedimentable RNase; *G, G′* soluble RNase; *H, H′* soluble proteinase. From BALZ (1966).

ethylene is the induction of flower fading which in *Ipomoea tricolor* is initiated only 90 minutes after exposure of newly opened flowers to ethylene (KENDE and BAUMGARTNER 1974); apart from the premature curling-up of the corolla, exogenous ethylene also advances the marked increase in RNase which characterizes senescence in this organ. In addition, KENDE and BAUM-GARTNER (1974) demonstrated that the normal fading and concomitant increase of RNase coincide with a sharp increase in the rate of endogenous ethylene production.

It should be mentioned that ethylene is an outstanding phytohormone in that it is not only a product of plant metabolism but a strong air pollutant as well. In early reports on the induction of senescence caused by leaking illuminating gas, the effects were attributed to ethylene, one of its minor constitutents. Since levels as low as 0.1 ppm may promote senescence (and abscission) of leaves, the accumulation of ethylene in the air which is caused by the combustion of fossil fuels is a serious potential hazard for the vegetation in highly industrialized and densely populated areas. ABELES (1973) reported on several cases of damage to crops caused by this particular form of air pollution.

In the explanted bean endocarp, ethylene significantly accelerates the lowering of protein levels, whereas only a small effect on the degradation of RNA has been observed (ABELES et al. 1967). On the other hand, auxin reduces the rate of senescence in this tissue (SACHER and SALMINEN 1969); ABELES (1973) discussed the possible mutual effects of auxin and ethylene in the control of senescence.

Mechanical damage to leaf and tuber tissues of potato causes a marked increase in RNase and, to a minor extent, in acid phosphatase and phosphodiesterase (PITT and GALPIN 1971). Similar responses to tissue damage have been observed in other objects. It is tempting to speculate that these effects are ultimately caused by ethylene. Effects of mechanical damage should perhaps be compared with the consequences of infection by fungi and bacteria; it seems to be a common observation that infection causes considerable enhancement of the production of ethylene as compared to the production in the healthy tissue (see HISLOP et al. 1973). This may be provoked by ethylene production in cells ruptured upon the initial phase of invasion. Of the various effects of ethylene which could be significant with regard to the resistance of host plants, the induced production of cell wall lytic enzymes (see p. 129) is particularly interesting. In Phaseolus leaves ethylene promotes the synthesis of β-1,3-glucanase and chitinase which may constitute a defense mechanism against fungal invasion (ABELES et al. 1970).

Returning to the lytic phenomena associated with senescence, reference must be made to the hormonal regulation of hydrolases which are involved in the lysis of structural polysaccharides. Excision of petiole abscission zones of Phaseolus vulgaris resulted in an increase of endo-β-1,3-glucanase (ABELES and FORRENCE 1970); cytokinins, auxin and gibberellic acid prevented the increase of this hydrolase whereas ethylene promoted it. A correlation between the amount of callose present in sieve plates and the enzyme levels suggests that β-1,3-glucanase is involved in the degradation of callose. In leaves of Nicotiana glutinosa senescence is characterized by the marked enhancement of endo-β-1,3-glucanase activity (MOORE and STONE 1972); abscisic acid which accelerates the loss of chlorophyll and protein causes, however, a reduction of the rate and extent of rise in this hydrolase. It is, therefore, difficult to attribute a specific function in senescence to β-1,3-glucanase. MITTELHEUSER and VAN STEVENINCK (1972) observed, in electron micrographs of senescent wheat leaf cells, that cell walls become thinner as senescence proceeds. This suggests that ageing of leaves is associated with a partial lysis of cell walls. Indeed, a conspicuous loss of noncellulosic polysaccharides occurs in discs of tobacco leaves. If they are floated on a cytokinin solution this loss is delayed, whereas in discs treated with abscisic acid it is somewhat accelerated (MATILE 1974). It has not been possible, however, to correlate these effects of hormones with the levels of glycosidases.

The most convincing correlations between a lytic phenomenon and hydrolase levels controlled by hormones have been demonstrated in the case of abscission of lateral organs. During senescence of the explanted pulvinar

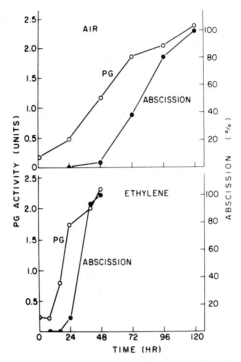

Fig. 53. Time course of the development of polygalacturonase activity and abscission of *Citrus sinensis* leaf explants incubated either in air (upper) or in 10 ppm ethylene (lower). From RIOV (1974).

tissue of *Phaseolus* leaves the increase in cellulase activity in the separation zone appears to be initiated by an increase in ethylene production; the application of abscisic acid advances the peak of ethylene production, the induction of the rise in cellulase activity and abscission, treatment with auxin retards all these (JACKSON and OSBORNE 1972). Hence, ethylene appears to be the factor which ultimately controls abscission. However, the effect of ethylene seems to depend also upon the levels of other hormones such as auxin and cytokinin which determine the sensitivity of the separation tissue to ethylene (see ABELES 1973).

In the context of the lytic compartment it is important to note that ethylene not merely raises the level of cellulase but also augments its secretion which, in conjunction with the production of polygalacturonase (Fig. 53) and perhaps other cell wall lytic enzymes, induces the separation of cells in abscission

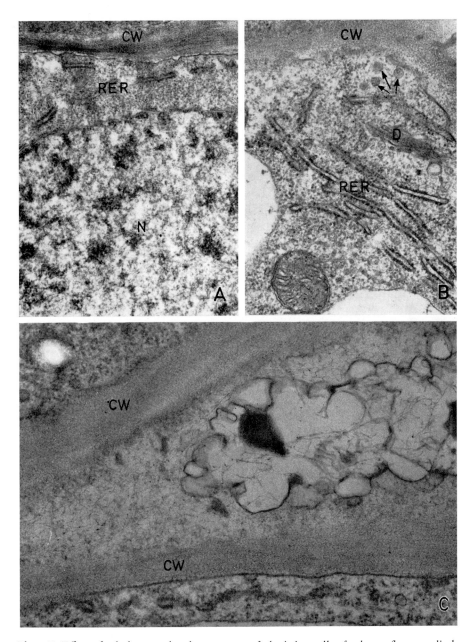

Plate 40. Effect of ethylene on the ultrastructure of abscission cells of tobacco flower pedicels.

A: Portion of a cortical cell of the abscission layer which has been treated with ethylene for 2 hours. ×31,000. *B:* After 5 hours of treatment with ethylene rough surfaced endoplasmic reticulum has accumulated. Golgi vesicles are denoted by the arrows. ×44,500. *C:* View of the wall regions of cortical cells after 5 hours treatment with ethylene. Note the extensive dissolution of the middle lamella. ×53,000. Courtesy of J. G. VALDOVINOS.

layers (*e.g.*, ABELES and LEATHER 1971, RIOV 1974). The necessary involvement of secretion, that is, transport of polysaccharidases synthesized in the cytoplasm into the walls of the cells of the separation layer, suggests that there is control at several levels: one level is certainly protein synthesis, a second level concerns membrane processes engaged in secretion and a third level of control concerns the necessary vectorial secretion of wall modifying enzymes into transverse walls of only the thin layer of abscission cells. According to VALDOVINOS *et al.* (1972), rough surfaced ER and dictyosomes

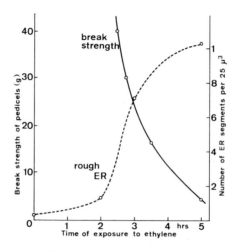

Fig. 54. Abscission of flower pedicels of tobacco plants. Relationship of break strength of the abscission zone and proliferation of rough ER in abscission cells to duration of exposure of plants to 5 ppm ethylene. Adapted from VALDOVINOS *et al.* (1972).

appear in cortical cells of abscission zones of tobacco flower pedicels in response to exposure to ethylene (Plate 40). The number of rough surfaced ER segments which can be observed in thin sections of this tissue in an area 25 μm^2 increases from 0.3 before treatment with ethylene to over 10 after 5 hours under treatment. This ethylene induced proliferation of the reticulum precedes the onset of cell separation as determined by recording the decrease of the force necessary to break the pedicel (Fig. 54). Whether these structural changes in the separation zone of tobacco pedicels are associated with the synthesis and eventual secretion of cellulase and other cell wall lytic enzymes is unknown. One is tempted, however, to assume that VALDOVINOS *et al.* (1972) detected the control of lysis at the level of cellular membrane systems. By analogy to abscission zones of other plants, which have been investigated for control at the biochemical level, it is likely that the rough ER is involved in the synthesis of polysaccharidases, the dictyosomes in the secretion of these enzymes. In any case, it will be difficult to obtain the necessary information about the fine structural localization of relevant hydrolases experimentally in order to establish the sequence of events.

4.4. Control of Hydrolase Production in Fungi

Under laboratory conditions the production and release of hydrolases in saprophytes appears to be regulated according to nutritional requirements. A wood decomposing fungus which is grown on a defined medium will secrete cellulase if cellulose is provided as the source of carbon; if glucose is the source of carbon, cellulase production will be small or even lacking. It would be incorrect to conclude from this common behaviour of saprophytic fungi that

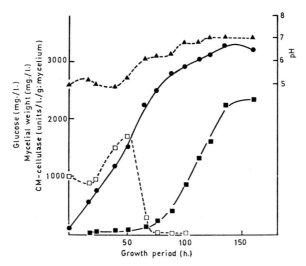

Fig. 55. Cellulase production in *Myrothecium verrucaria*. The culture medium originally contained 0.1% glucose; after 10 hours the 3.3 l fermentor culture was continuously supplied with fresh medium containing 4% glucose at a dilution rate of 0.3% per hour. ●——● mycelial dry weight; ■——■ cellulase activity per gram mycelium; □——□ glucose concentration in the medium; ▲——▲ pH. From HALUNE and STRANKS (1971).

the production of a specific hydrolase is induced in response to the presence of cellulose. *Myrothecium verrucaria* can produce cellulose-depolymerizing enzymes in the absence of cellulose or even in the absence of constitutents of cellulose, e.g., if grown on glycerol as a source of carbon (HULME and STRANKS 1971); the initiation of cellulase production coincides with the deceleration of mycelial growth when the source of carbon in the medium is largely exhausted (Fig. 55). Since in *Myrothecium* cellulase may be produced in the absence of cellulose under nutritional conditions that result in a reduced growth rate, it seems that this enzyme is constitutive and that synthesis may be subjected to *catabolite repression*. The existence of this type of control is rather surprising if it is considered that under natural conditions saprophytic fungi may always be exposed to some sort of starvation.

Hence, if any condition causing a reduced growth rate, that is, the limitation of any essential nutrient will derepress cellulase production, the control which can be detected in the laboratory would perhaps never be brought into

play under natural conditions. Should it be the case that the limitation of
growth, rather than the specific factor responsible for starvation, causes
derepression, then catabolite repression is perhaps only part of a more complex
regulation of hydrolase production.

Experiments on the control of proteinase secretion in moulds suggest,
indeed, that specific factors may be involved. *Neurospora crassa* secretes
proteinases in the presence of proteinaceous sources of nitrogen such as casein
or gelatine (Fig. 56). If an inorganic source of nitrogen such as ammonium,

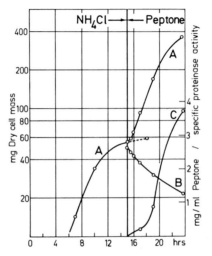

Fig. 56. Utilization of exogenous protein in *Neurospora crassa*. Secretion of acid proteinase
in the presence of peptone. *A* dry mass of mycelium; *B* concentration of peptone in the
medium; *C* released proteinase activity per unit of dry cell mass. Conidia (7×10^5/ml)
were inoculated into the initial medium containing a low concentration of ammonium
which was exhausted after 15 hours. At this time point fresh medium containing peptone
was added. After MATILE (1965).

nitrate or amino acids is provided together with the protein, secretion is
reduced (MATILE 1965). At concentrations of ammonium or amino acids
which allow a maximal growth rate of the mycelium, proteinase secretion may
be almost completely repressed. However, growth rates under derepressing
conditions, *e.g.*, in the presence of peptone as the sole source of nitrogen, may
be almost as high as in the presence of ammonium (Fig. 57). Results presented
by DRUCKER (1973) suggest that this phenomenon may be due to the presence
in peptone of small amounts of amino acids or other products of protein
degradation which are necessary to induce proteinase secretion: if mycelia of
Neurospora crassa in the exponential phase are transferred to a medium containing a highly purified protein source such as bovine serum albumin, the
mould starves because it produces very little extracellular proteolytic enzyme;
however, production can be induced by adding a proteinase (*e.g.*, a culture
filtrate from a mycelium induced to secrete proteinase to the medium). Hence,
the signal which induces proteinase secretion in *Neurospora* appears to be
given by a product of extracellular protein degradation (DRUCKER 1973).

Whether or not this product is identical with amino acids is unknown; in any case it must be absorbed by the hyphae in order to interact in some way with the protein synthesizing system. Upon induction, the extracellular proteinases are formed *de novo* (HEINIGER and MATILE 1974), which means that the exogenous factor triggers protein synthesis and not merely the secretion of preexisting enzymes. A possible role of small amounts of amino acids in the induction of proteinase secretion is also suggested by the finding that secretion occurs under various conditions of starvation. If growth of *Neurospora*

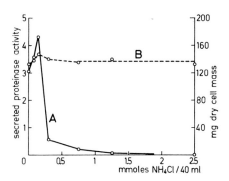

Fig. 57. Repression of proteinase secretion by ammonia in *Neurospora crassa*. The mycelium was cultured for 16 hours in the presence of peptone (2.62 mg/ml) and different concentrations of NH_4Cl. *A* proteinase activity per unit of dry cell mass. *B* dry cell mass. From MATILE (1965).

mycelia is limited in the absence of nitrogen, carbon or sulfur, secretion is induced (HEINIGER and MATILE 1974). This may be due to a limited autolysis and release of amino acids (or of cytoplasmic proteins and intracellular constitutive proteinases) which occurs in the oldest parts of the hyphae. Sulfur deficiency causing the induction of proteinase secretion has also been observed in *Aspergillus nidulans* (TOMANAGA *et al.* 1964). KLAPPER *et al.* (1973) interpreted the finding that in *Aspergillus oryzae* proteolytic activity begins to appear in the medium when the glucose is exhausted, in terms of catabolite repression of proteinase secretion. Whether, in fact, all of these effects concern a common mechanism of control of extracellular proteinase production in these moulds cannot be decided at the present time. It is tempting to hypothesize, however, that a very small concentration of amino acids is necessary for induction of proteinase secretion. According to this hypothesis induction would be specific but a variety of circumstances may be responsible for the appearance of inducer molecules in the environment of the mould. Limited autolysis of hyphae occurring under conditions of starvation could be a possible source of amino acids. However, high concentration of nitrogen sources, the utilization of which does not require proteolytic action, seem to repress the production of extracellular proteinases (Fig. 57).

Under natural conditions secreted phosphatases may be functionally significant in the hydrolysis of naturally occurring sources of phosphate. Baker's yeast can, indeed, be grown in the presence of phosphomonoesters such as glycerophosphate (GÜNTHER and KATTNER 1968). In addition, acid phosphatase activity is repressed in the presence of high concentrations of inorganic

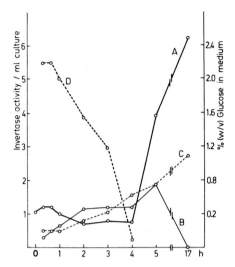

Fig. 58. Changes of activities of large and small invertase isoenzymes in the course of a culture cycle of *Saccharomyces* strain 303—367 on glucose. *A* secreted invertase; *B* internal large invertase; *C* internal small invertase; *D* glucose concentration in the culture medium. From MEYER and MATILE (1974 b).

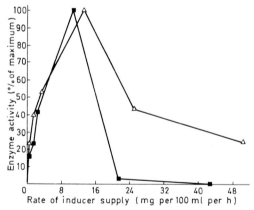

Fig. 59. Synthesis of endo-polygalacturonase (■———■) and arabinase (△———△) by *Verticillium albo-atrum* in restricted cultures supplied with increasing rates of galacturmic acid and arabinose, respectively. Enzyme activities are expressed as percentages of the maximum specific activity attained 10 hours after addition of inducer. From COOPER and WOOD (1973).

phosphate, derepressed under conditions of phosphate limitation (*e.g.*, SUOMALAINEN *et al.* 1960, SCHURR and YAGIL 1971).

Repression of phosphatases by orthophosphate have also been established in *Neurospora crassa* (see JACOBS 1973) as well as in other fungi, algae and even in a higher plant, the lemnaceae *Spirodela* (see REID and BIELESKI 1970).

The potential involvement of (external) phosphatases which are derepressed under conditions of phosphorus deficiency is indirectly suggested by the fact, that in *Neurospora crassa* the nucleases secreted, which would produce nucleotides from exogenous nucleic acids, are also repressed in the presence of orthophosphate (HASUNUMA 1973). Hence, it appears that phosphorus deficiency causes derepression of several external hydrolases which are potentially capable of producing resorbable phosphate from various natural phosphate compounds. It may be speculated that the production of these enzymes is regulated by the internal orthophosphate concentrations which in yeast

Table 6. *Induction of Synthesis of Polysaccharidases of Verticillium albo-atrum by the Restricted Supply of Cell Wall Sugars* (after COOPER and WOOD 1973).
Enzyme activities are expressed as a percentage of the maximum specific activities attained in culture after identical periods of exposure to different sugars. The sugars were fed from diffusion capsules at linear rates of 2–4 mg per 100 ml per hour to shake cultures containing initially 4×10^9 spores per ml

Inducer	Enzyme activities			
	Endo-poly-galacturonase	Arabinase	Xylanase	Cellulase
D-Galacturonic acid	**100**	15.8	16.5	0.30
L-Arabinose	0	**100**	3.5	0.25
D-Xylose	0	3.7	**100**	0.08
D-Glucose	0—1.25	0—5.8	5.8	1.46
Cellobiose	3.7	—	—	**100**

would become very low under conditions of phosphate deficiency because the cells produce a vacuolar store of polyphosphate. It is interesting to note that phosphate repression is as effective in yeast spheroplasts as in the intact cells (VAN RIJN et al. 1972). This is not the case in the regulation of external invertase levels. It is well documented that yeast invertase is subjected to repression by hexose (see LAMPEN et al. 1972); this control cannot be demonstrated in yeast spheroplasts; with concentrations of glucose in the medium as high as 5% the secretion of invertase is not repressed. In contrast, external invertase is not produced in a growing population of intact cells unless the glucose concentration in the medium has reached a very low threshold (Fig 58; MEYER and MATILE 1974 b).

Several cases of induction of secretion of hydrolases in the presence of specific inducers have been reported. The functional significance of the apparently specific induction of cellobiase by cellobiose in *Neurospora crassa* (EBERHART and BECK 1973) is not clear, However, the induction of glycosidases in response to the presence of the corresponding free sugars which has been detected in pathogenic fungi (*e.g.*, KEEN et al. 1970, KEEGSTRA et al. 1972, ENGLISH et al. 1971) is significant. It is likely that small amounts of free sugars are present in the cell walls of the host tissues as a result of cell wall degradation associated with growth and differentiation. If these sugars induce the secretion of the corresponding glycosidases in the invading fungus, an effective lysis of host cell walls must be the result. Moreover, the production of free sugars in the course of infection could explain the sequential induction of the various glycosidases which is typical for pathogenic fungi cultured on isolated host cell walls as a source of carbon (see p. 128). COOPER and WOOD (1973) have reported a marked dependency of glycosidase production in vascular wilt fungi on the concentration of the inducers. In *Verticillium albo-atrum* a high degree of specificity of induction of endopolygalacturonase by galacturonic acid (and other hydrolases by corresponding inducers) has been detected (Table 6). If galacturonic acid is supplied to cultures of this fungus at increasing rates, a high specific polygalacturonase activity is produced at a comparatively low rate whereas high rates repress the production completely (Fig. 59). This means that hydrolase secretion in *V. albo-atrum* is controlled specifically by inducing sugars and unspecifically by catabolite repression which comes into action when large quantities of metabolizable sugars are available. This brings us back to the beginning of this section. Regardless of the nutritional conditions of both saprophytic and parasitic fungi some kind of starvation seems to be a most important factor in the control of hydrolase production.

Retrospect: "Life's Irreducible Structure"

The laboratories of biochemists are usually supplied with very complicated metabolic charts that remind one of the wiring diagrams of electronic devices. It is remarkable that most of the enzymes which play a prominent role in the present monograph are not considered in these charts. The reason is simply that it would be impossible to integrate the unspecific hydrolases in a two-dimensional scheme of metabolism. The integration of these enzymes requires the introduction of an additional dimension, that of compartmentation.

Admittedly, it is a truism that most of the enzyme reactions which appear on the biochemical charts, either take place in distinct cell compartments or are integrated in multienzyme complexes and cellular membranes. Biological research of the past decades has produced an overwhelming amount of facts which demonstrate the importance of cellular structures in the organization and control of metabolism. Yet none of the structure/function relationships reveals as strikingly as does the lytic cell compartment that an entire dimension is missing on the biochemical charts. Without the membranes of this compartment which separate the unspecific hydrolases from the sites of anabolic reactions, a living cell can hardly be visualized.

It is interesting to note that in the theory of life that has emerged from the activity of molecular biologists the structural dimension of the cell is not adequately considered. As the aim of research on the inheritance of biological specificity was to find an explanation of this phenomenon in biochemical terms, it is not surprising that the theory is based exclusively on the functions of the biopolymers DNA, RNA and protein. The "grammar of life" concerns the information which is inscribed in DNA, and the theory of life comprises largely the impact of this information with regard to the specificity of proteins, which subsequently determine the specificity of metabolic reactions and, hence, the biological specificity as a whole.

There is no doubt that the ingenuity of molecular biologists has elucidated a decisive living mechanism. However, in my opinion the theory is incomplete in that it is restricted to the biochemical aspects of life. To quote one of the pioneers of nucleic acid biochemistry, CHARGAFF (1971): "Somehow I cannot rid myself of the feeling that we still lack an entire dimension that is necessary for the understanding of a living cell". This dimension which is lacking in the biochemical theory of life is nothing less than the cell itself. In fact, the continuous interaction of biochemical mechanisms is completely dependent on their integration in living cells. None of the regulatory systems can be described in biochemical terms exclusively; in all cases structural entities of the cell are intimately involved in the organization of the biochemical machinery. *In vitro* it is only possible to demonstrate fractions cut out of the metabolic complex and this, only provided the conditions selected imitate the conditions in the living cell. In the case of the lytic compartment this is vividly illustrated by the fact that nucleases must be removed from preparations of nucleic acids which are used in experiments of nucleic acid and protein synthesis *in vitro*. In other words, the biochemist has to separate components which in the living cell are localized in distinct compartments. Hence, in my opinion the cell represents a principle in its own right. An adequate theory of life would, therefore, have to take into account that this principle is irreducible to the level of biochemical mechanisms. As pointed out by POLANY (1968), the structure of life is characterized by a whole hierarchy of controlling principles. All of these priciples represent boundary conditions harnessing the laws at lower levels of biological activity. At the level of the cell such boundary conditions can be described morphologically, for instance, in the case of the lytic compartment, in terms of membranes which separate the cellular digestive processes from other branches of metabolism, but they cannot be reduced to the biochemical level associated with information in DNA. I think that evolution of biological sciences has now reached the point at which "life's irreducible structure" (POLANY) becomes perceivable.

Bibliography

ABELES, F. B., 1969: Abscission: role of cellulase. Plant Physiol. **44**, 447—452.

— 1973: Ethylene in plant biology. New York-London: Academic Press.

— R. P. BOSSHART, L. E. FORRENCE, and W. H. HABIG, 1970: Preparation and purification of glucanase and chitinase from bean leaves. Plant Physiol. **47**, 129—134.

— and L. E. FORRENCE, 1970: Temporal and hormonal control of β-1,3-glucanase in *Phaseolus vulgaris* L. F. B. Plant Physiol. **45**, 395—400.

— R. E. HOLM, and H. E. GAHAGAN, 1967: Abscission: the role of ageing. Plant Physiol. **42**, 1351—1356.

— and G. R. LEATHER, 1971: Abscission: control of cellulase secretion by ethylene. Planta **97**, 87—91.

ABU-ZINADA, A.-A. H., A. COBB, and D. BOULTER, 1973: Fine structural studies on the infection of *Vicia faba* L. with *Botrytis fabae* Sard. Arch. Microbiol. **91**, 55—66.

ACTON, G. J., 1972: Phytochrome control of the level of extractable ribonuclease activity in etiolated hypocotyls. Nature-New Biol. **236**, 255—256.

ALBERSHEIM, P., T. M. JONES, and P. D. ENGLISH, 1969: Biochemistry of the cell wall in relation to infective processes. Ann. Rev. Phytopathol. **7**, 171—194.

— and B. S. VALENT, 1974: Host-pathogen interactions. VII. Plant pathogens secrete proteins which inhibit enzymes of the host capable of attacking the pathogen. Plant Physiol. **53**, 684—687.

ALLEN, C. F., P. GOOD, H. H. MOLLENHAUER, and C. TOTTEN, 1971: Studies on seeds. IV. Lipid composition of bean cotyledon vesicles. J. Cell Biol. **48**, 542—546.

AMAGASE, S., 1972: Digestive enzymes in insectivorous plants. III. Acid proteases in the genus *Nepenthes* and *Drosera peltata*. J. Biochem. Tokyo **72**, 73—81.

ANDERSON, A. J., and P. ALBERSHEIM, 1972: Host-pathogen interactions. V. Comparison of the abilities of proteins isolated from three varieties of *Phaseolus vulgaris* to inhibit the endopolygalacturonase secreted by three races of *Colletotrichum lindemuthianum*. Physiol. Plant Pathol. **2**, 339—346.

ANDERSON, J. W., and K. S. ROWAN, 1966: The effect of 6-furfurylaminopurine on senescence in tobacco-leaf tissue after harvest. Biochem. J. **98**, 401—404.

ARAD, S. M., Y. MIZRAHI, and A. E. RICHMOND, 1973: Leaf water content and hormone effects on ribonuclease activity. Plant Physiol. **52**, 510—512.

ARMENTROUT, V. N., G. G. SMITH, and C. L. WILSON, 1968: Spherosomes and mitochondria in the living fungal cell. Amer. J. Bot. **55**, 1062—1067.

ARMSTRONG, J. E., and R. L. JONES, 1973: Osmotic regulation of α-amylase synthesis and polyribosome formation in aleurone cells of barley. J. Cell Biol. **59**, 444—455.

ASHFORD, A. E., 1970: Histochemical localization of β-glucosidase in roots of *Zea mays*. I. A simultaneous coupling azo-dye technique for the localization of β-glucosidase and β-galactosidase. Protoplasma **71**, 281—293.

— and J. V. JACOBSEN, 1974: Cytochemical localization of phosphatase in barley aleurone tissue: the pathway of gibberellic acid induced enzyme release. Planta **120**, 81—106.

— and M. E. McCULLY, 1970 a: Histochemical localization of β-glucosidases in roots of *Zea mays*. II. Changes in localization and activity of β-glucosidase in the main root apex. Protoplasma **71**, 389—402.

— — 1970 b: Localization of naphthol AS-B 1 phosphatase activity in lateral and main root meristems of pea and corn. Protoplasma **70**, 441—456.

ATKIN, R. K., and B. I. S. SRIVASTAVA, 1969: The changes in soluble protein of excised barley leaves during senescence and kinetin treatment. Physiol. Plant. **22**, 742—750.

AVERS, C. J., and E. E. KING, 1900: Histochemical evidence of intracellular enzymatic heterogeneity of plant mitochondria. Amer. J. Bot. **47**, 220—225.

BAILEY, C. J., A. COBB, and A. BOULTER, 1970: A cotyledon slice system for the electron autoradiographic study of the synthesis and intracellular transport of the seed storage protein of *Vicia faba*. Planta **95**, 103—118.

BAIN, J. M., and F. V. MERCER, 1966: Subcellular organization of the cotyledons in germinating seeds and seedlings of *Pisum sativum* L. Aust. J. Biol. Sci. **19**, 69—84.

BAL, A. K., and J. F. PAYNE, 1972: Endoplasmic reticulum activity and cell wall breakdown in quiescent root meristems of *Allium cepa* L. Z. Pflanzenphysiol. **66**, 265—272.

BALZ, H. P., 1966: Intrazelluläre Lokalisation und Funktion von hydrolytischen Enzymen bei Tabak. Planta **70**, 207—236.

BARCKHAUS, R., 1973: Zur Vakuolenbildung in der sich differenzierenden Pflanzenzelle. Naturwissenschaften **60**, 302—303.

BARRAS, D. R., 1972: A β-glucan endo-hydrolase from *Schizosaccharomyces pombe* and its role in cell wall growth. Anton Leeuwenhoek J. Microbiol. **38**, 65—80.

BARTNICKI-GARCIA, S., and E. LIPPMAN, 1972: The bursting tendency of hyphal tips of fungi: presumptive evidence for a delicate balance between wall synthesis and wall lysis in apical growth. J. Gen. Microbiol. **73**, 487—500.

BARTON, R., 1965: Electron microscopic studies on the origin and development of the vacuoles in root tip cells of *Phaseolus*. Cytologia (Tokyo) **30**, 266—273.

BATEMAN, D. F., and R. L. MILLAR, 1966: Pectic enzymes in tissue degradation. Annu. Rev. Phytopathol. **4**, 119—146.

BAUER, H., and E. SIGARLAKIE, 1973: Cytochemistry on ultrathin frozen sections of yeast cells. J. Microsc. **99**, 205—218.

BAUM, H., and K. S. DODGSON, 1957: Differentiation between myrosulphatase and the arylsulphatases. Nature **179**, 312—313.

BAUMGARTNER, B., H. KENDE, and PH. MATILE, 1974: RNase in ageing Japanese morning glory. Plant Physiol. (in press).

BAUR, P. S., and C. H. WALKINSHAW, 1974: Fine structure of tannin accumulations in callus cultures of *Pinus elliotti*. Can. J. Bot. **52**, 615—620.

BECK, C., and H. K. VON MEYENBURG, 1968: Enzyme pattern and aerobic growth of *Saccharomyces cerevisiae* and various degrees of glucose limitation. J. Bacteriol. **96**, 479—486.

BEEVERS, H., 1969: Glyoxysomes of castor bean endosperm and their relation to gluconeogenesis. Ann. N.Y. Acad. Sci. **168**, 313—324.

BEEVERS, L., 1908: Protein degradation and proteolytic activity in the cotyledons of germinating pea seeds. Phytochemistry **7**, 1837—1844.

BENNETT, P. A., and M. J. CHRISPEELS, 1972: *De novo* synthesis of ribonuclease and β-1,3-glucanase by aleurone cells of barley. Plant Physiol. **49**, 445—447.

BERJAK, P., 1968: A lysosome-like organelle in the root cap of *Zea mays*. J. Ultrastruct. Res. **23**, 233—242.

— 1972: Lysosomal compartmentation: ultrastructural aspects of the origin, development, and function of vacuoles in root cells of *Lepidium sativum*. Ann. Bot **36**, 73—81.

— and J. R. LAWTON, 1973: Prostelar autolysis: a further example of a programmed senescence. New Phytol. **72**, 625—637.

— and T. A. VILLIERS, 1970: Ageing in plant embryos. I. The establishment of the sequence of development and senescence in the root cap during germination. New Phytol. **69**, 929—938.

— — 1972: Ageing in plant embryos. V. Lysis of the cytoplasm in non-viable embryos. New Phytol. **71**, 1075—1079.

BERTINI, F., D. BRANDES, and D. E. BUETOW, 1965: Increased acid hydrolase activity during carbon starvation in *Euglena gracilis*. Biochim. biophys. Acta **107**, 171—173.

BESFORD, R. T., and G. E. HOBSON, 1972: Pectic enzymes associated with the softening of tomato fruit. Phytochemistry **11**, 2201—2205.

BETETA, P., and S. GASCÓN, 1971: Localization of invertase in yeast vacuoles. FEBS lett. **13**, 297—300.

BETINA, V., D. MIČEKOVÁ, and P. NEMEC, 1972: Antimicrobial properties of cytochalasins and their alteration of fungal morphology. J. Gen. Microbiol. **71**, 343—349.

BIELY, P., V. FARKAŠ, and Š. BAUER, 1972: Secretion of β-glucanase by *Saccharomyces cerevisiae* protoplasts. FEBS lett. **23**, 153—156.

— Z. KRÁTKÝ, J. KOVAŘÍK, and Š. BAUER, 1971: Effect of 2-deoxyglucose on cell wall formation in *Saccharomyces cerevisiae* and its relation to cell growth inhibition. J. Bacteriol. **107**, 121—129.

BIGGER, C. H., M. R. WHITE, and H. D. BRAYMER, 1972: Ultrastructure and invertase secretion of the slime mutant of *Neurospora crassa*. J. Gen. Microbiol. **71**, 159—166.

BLUM, J. J., and D. E. BUETON, 1963: Biochemical changes during acetate deprivation and repletion in *Euglena*. Exp. Cell Res. **29**, 407—421.

BOGEN, H. J., and M. KESER, 1954: Eiweißabbau durch Acridinorange bei Hefezellen. Physiol. Plant **7**, 446—462.

BORNMAN, C. H., 1967: Some ultrastructural aspects of abscission in *Coleus* and *Gossypium*. S. Afr. J. Sci. **63**, 325—331.

BOWES, B. G., 1965: The origin and development of vacuoles in *Glechoma hederacea* L. La Cellule **65**, 359—364.

— 1969: Electron microscopic observations on myelin-like bodies and related membranous elements in *Glechoma hederacea* L. Z. Pflanzenphysiol. **60**, 414—417.

BOYLEN, CH. W., and CH. W. HAGEN, 1969: Partial purification and characterization of a flavonoid-3-β-D-glucosidase from petals of *Impatiens balsamina*. Phytochemistry **8**, 2311—2315.

BRACKER, CH. E., 1971: Cytoplasmic vesicles in germinating spores of *Gilbertella persicaria*. Protoplasma **72**, 381—397.

BRANDES, D., and F. BERTINI, 1964: Role of Golgi apparatus in the formation of cytolysomes. Exp. Cell Res. **35**, 194—217.

BRAUN, A. C., and H. N. WOOD, 1966: On the inhibition of tumor inception in the crown gall disease with the use of ribonuclease. Proc. nat. Acad. Sci. (US) **56**, 1417—1422.

BRIARTY, L. G., D. A. COULT, and D. BOULTER, 1969: Protein bodies of developing seeds of *Vicia faba*. J. Exp. Bot. **20**, 358—372.

— — — 1970: Protein bodies of germinating seeds of *Vicia faba*. Changes in fine structure and biochemistry. J. Exp. Bot. **21**, 513—524.

BURR, F. A., and J. A. WEST, 1971: Protein bodies in *Bryopsis hypnoides:* their relationship to wound-healing and branch septum development. J. Ultrastruct. Res. **35**, 476—498.

BUSHNELL, W. R., 1972: Physiology of fungal haustoria. Ann. Rev. Phytopathol. **10**, 151—176.

BUTLER, R. D., 1967: The fine structure of senescing cotyledons of cucumber. J. Exp. Bot. **18**, 535—543.

BUTTROSE, M. S., 1963: Ultrastructure of the developing aleurone cells of wheat grain. Aust. J. Biol. Sci. **16**, 768—774.

BUVAT, R., 1957: Relations entre l'ergastoplasme et l'appareil vacuolaire. C. R. Acad. Sci. (Paris) **245**, 350—352.

— 1968: Diversité des vacuoles dans les cellules de la racine d'orge (*Hordeum sativum*). C. R. Acad. Sci. (Paris) **267**, 296—298.

— et A. MOUSSEAU, 1960: Origine et évolution du système vacuolaire dans a racine de « *Triticum vulgare* » ; relation avec l'ergastoplasme. C. R. Acad. Sci. (Paris) **251**, 3051—3053.

CABIB, E., and R. ULANE, 1973: Chitin synthetase activating factor from yeast, a protease. Biochem. biophys. Res. Commun. **50**, 186—191.

CALONGE, F. D., A. H. FIELDING, R. J. W. BYRDE, and O. A. AKINREFON, 1969: Changes in ultrastructure following fungal invasion and the possible relevance of extracellular enzymes. J. Exp. Bot. **20**, 350—357.

CARLSON, P. S., 1972: Notes on the mechanism of action of gibberellic acid. Nature-New Biol. **237**, 39—41.

CARROLL, F. E., and G. C. CARROLL, 1973: Senescence and death of the conidiogenous cell in *Stemphylium botryosum* Wallroth. Arch. Mikrobiol. **94**, 109—124.

CATESSON, A.-M., et Y. CZANINSKI, 1967: Mise en évidence d'une activité phosphatasique acide dans le réticulum endoplasmique des tissus conducteurs de robinier et de sycomore. J. Microsc. (Paris) **6**, 509—514.

— — 1968: Localisation ultrastructurale de la phosphatase acide et cycle saisonnier dans les tissus conducteurs de quelques arbres. Bull. Soc. Franç. Physiol. Végét. **14**, 165—173.

— R. GOLBERG et M.-C. WINNY, 1971: Etude d'activités phosphatasiques acides dans les cellules d'*Acer pseudoplatanus* cultivées en suspension. C. R. Acad. Sci. (Paris) **272**, 2078—2081.

CHAFE, S. C., and D. J. DURZAN, 1973: Tannin inclusions in cell suspension cultures of white spruce. Planta **113**, 251—262.

CHAKRAVORTY, A. K., and M. SHAW, 1971: Changes in the transcription pattern of flax cotyledons after inoculation with flax rust. Biochem. J. **123**, 551—557.

— — and L. A. SCRUBB, 1974: Ribonuclease activity of wheat leaves and rust infection. Nature **247**, 577—580.

CHAN, Y.-H., and W. E. SACKSTON, 1972: Production of pectolytic and cellulolytic enzymes by virulent and avirulent isolates of *Sclerotium bataticola* during disease development in sunflowers. Can. J. Bot. **50**, 2449—2453.

CHANG, P. L. Y., and J. R. TREVITHICK, 1970: Biochemical and histochemical localization of invertase in *Neurospora crassa* during conidial germination and hyphal growth. J. Bacteriol **102**, 423—429.

— — 1972 a: Distribution of wall-bound invertase during the asexual life-cycle of *Neurospora crassa*. J. Gen. Microbiol. **70**, 23—29.

— — 1972 b: Release of wall-bound invertase and trehalase in *Neurospora crassa* by hydrolytic enzymes. J. Gen. Microbiol. **70**, 13—22.

— — 1974: How important is secretion of exoenzymes through apical cell walls of fungi? Arch. Microbiol. **101**, 281—294.

CHARGAFF, E., 1971: Preface to a grammar of biology. Science **172**, 637—642.

CHEN, A. W., and J. J. MILLER, 1968: Proteolytic activity of intact yeast cells during sporulation. Can. J. Microbiol. **14**, 957—963.

CHIN, T. Y., R. POULSON, and L. BEEVERS, 1972: The influence of axis removal on protein metabolism in cotyledons of *Pisum sativum* L. Plant Physiol. **49**, 428—489.

CHING, T. M., 1968: Ultracellular distribution of lipolytic activity in the female gameto-phyte of germinating Douglas fir seeds. Lipids **3**, 482—488.

— 1970: Glyoxysomes in megagametophyte of germinating ponderosa pine seeds. Plant Physiol. **46**, 475—482.

CHRISPEELS, M. J., and D. BOULTER, 1974: Exhancement of endopeptidase activity in the cotyledons of germinating mungbeans. Plant Physiol. (in press).

— and J. E. VARNER, 1967: Gibberellic acid-enhanced synthesis and release of α-amylase and ribonuclease by isolated barley aleurone layers. Plant Physiol. **42**, 398—406.

CLAES, H., 1971: Autolyse der Zellwand bei den Gameten von *Chlamydomonas reinhardii*. Arch. Mikrobiol. **78**, 180—188.

CLARKE, A. E., and B. A. STONE, 1962: β-1,3-glucan hydrolase from the grape vine (*Vitis vinifera*) and other plants. Phytochemistry **1**, 175—188.

COCKING, E. C., 1960: A method for the isolation of plant protoplasts and vacuoles. Nature **187**, 962—963.

COOK, B. E., 1972: The distribution and role of cytoplasmic membrane-bounded vesicles during the development of botryose solitary blastospores of two fungi. New Phytol. **71**, 1135—1141.

COOPER, R. M., and R. K. S. WOOD, 1973: Induction of cell wall degrading enzymes in vascular wilt fungi. Nature **246**, 309—310.

CORBETT, J. R., and C. A. PRICE, 1967: Intracellular distribution of p-nitrophenylphosphatase in plants. Plant Physiol. **42**, 827—830.

CORTAT, M., PH. MATILE, and F. KOPP, 1973: Intracellular localization of mannan synthetase activity in budding baker's yeast. Biochem. biophys. Res. Commun. **53**, 482—489.

— — and A. WIEMKEN, 1972: Isolation of glucanase-containing vesicles from budding yeast. Arch. Mikrobiol. **82**, 189—205.

COULOMB, C., 1973 a: Phénomènes d'autophagie liés à la différentiation cellulaire dans les jeunes recines de scorsonère (*Scorzonera hispanica*). C. R. Acad. Sci. (Paris) **276**, 1161—1164.

— 1973 b: Diversité des corps multivésiculaires et notion d'hétérophagie dans le méristème radiculaire de scorsonère (*Scorzonera hispanica*). J. Microsc. (Paris) **16**, 345—360.

— et R. BUVAT, 1968: Processus de dégénérance cytoplasmique partielle dans les cellules de jeunes racines de *Cucurbita pepo*. C. R. Acad. Sci. (Paris) **267**, 843—844.

— et PH. COULOMB, 1973: Participation des structure golgiennes à la formation des vacuoles autolytiques et à leur approvisionnement enzymatique dans les cellules du méristème radiculaire de la courge. C. R. Acad. Sci. (Paris) **277**, 2685—2688.

COULOMB, PH., 1968: Etude préliminaire sur l'activité phosphatasique acide des particules analogues aux lysosomes des cellules radiculaires jeunes de la courge (*Cucurbita pepo* L. Cucurbitacée). C. R. Acad. Sci. (Paris) **267**, 2133—2136.

— 1969: Localisation de la déoxyribonucléase acide dans le méristème radiculaire de la courge (*Cucurbita pepo* L. Cucurbitacée). C. R. Acad. Sci. (Paris) **269**, 1514—1516.

— 1971 a: Phytolysosomes dans le méristème radiculaire de la courge (*Cucurbita pepo* L. Cucurbitacée). Activité phosphatasique acide et activité peroxydasique. C. R. Acad. Sci. (Paris) **272**, 42—51.

— 1971 b: Phytolysosomes dans les frondes d'*Asplenium fontanum* (Filicinées, Poly-podiacées). Isolement sur gradient dosages de quelques hydrolases et contrôle des culots obtenus, par la microscopie électronique. J. Microsc. (Paris) **11**, 299—318.

— et C. COULOMB, 1972: Processus d'autophagie cellulaire dans les cellules de méristèmes radiculaires en état d'anoxie. C. R. Acad. Sci. (Paris) **274**, 214—217.

— — et J. COULON, 1972: Origine et fonctions des phytolysosomes dans le méristème radiculaire de la courge (*Cucurbita pepo* L. Cucurbitacée). I. Origine des phytolysosomes. Relations reticulum endoplasmique — dictyosomes — phytolysosomes. J. Microsc. (Paris) **13**, 263—280.

— et J. COULON, 1971: Fonctions de l'appareil de Golgi dans les méristèmes radiculaires de la courge (*Cucurbita pepo* L. Cucurbitacée). J. Microsc. (Paris) **10**, 203—214.

— — 1971 b: Sites de synthèse et migration des protéines dans les cellules du méristème radiculaire de la courge *(Cucurbita pepo* L. Cucurbitacée). Formation des phytolysosomes. Etude autoradiographique en microscopie électronique. C. R. Acad. Sci. (Paris) **272**, 1757—1759.

COUPÉ, M., S. PUJARNISCLE et J. D'AUZAC, 1972: Compartimentation de diverses oxydo-réductases (peroxydase, o-diphénol oxydase et malate déshydrogénase) dans le latex d'*Hevea brasiliensis* (Kunth), Müll. Arg. Physiol. Végét. **10**, 459—480.

CRESTI, M., E. PACINI, and G. SARFATTI, 1972: Ultrastructural studies on the autophagic vacuoles in *Eranthis hiemalis* endosperm. J. Submicr. Cytol. **4**, 33—44.

CRONSHAW, J., and J. B. BOUCK, 1965: The fine structure of differentiating xylem elements. J. Cell Biol. **24**, 415—431.

— and I. CHARVAT, 1973: Localization of β-glycerophosphatase activity in the myxomycete *Perichaena vernicularis*. Can. J. Bot. **51**, 97—101.

CZANINSKI, Y., 1968: Etude cytologique de la différenciation cellulaire du bois de robinier. I. Différenciation des vaisseaux. J. Microsc. (Paris) **7**, 1051—1068.

— 1972: Observations ultrastructurales sur l'hydrolyse des parois primaires des vaisseaux chez *Robinia pseudacacia* L. et *Acer pseudoplatanus*. C. R. Acad. Sci. (Paris) **275**, 361—364.

— and A.-M. CATESSON, 1970: Activités peroxydasiques d'origines diverses dans les cellules d'*Acer pseudoplatanus* (Tissues conducteurs et cellules en culture). J. Microsc. (Paris) **9**, 1089—1102.

DATKO, A. H., and G. A. MacLACHLAN, 1968: IAA and the synthesis of glucanases and pectic enzymes. Plant Physiol. **43**, 735—742.

DAUWALDER, M., W. G. WHALEY, and J. E. KEPHART, 1969: Phosphatases and differentiation of the Golgi apparatus. J. Cell Sci. **4**, 455—497.

DAVENPORT, T. I., and N. G. MARINOS, 1971: Cell separation in isolated abscission zones. Aust. J. Biol. Sci. **24**, 709—715.

Davies, E., and G. A. MacLachlan, 1968: Effects of indoleacetic acid on intracellular distribution of β-glucanase activities in the pea epicotyl. Arch. biochem. biophys. **128**, 595—600.
— 1969: Generation of cellulase activity during protein synthesis by pea microsomes *in vitro*. Arch. biochem. biophys. **129**, 581—587.
de Duve, Ch., 1969: The lysosomes in retrospect. In: Lysosomes in biology and pathology (Dingle, J. T., and H. B. Fell, eds.), Vol. 1, pp. 3—40. Amsterdam-London: North-Holland Publishing Company.
de Leo, P., and J. A. Sacher, 1970: Control of Ribonuclease and abscisic acid during senescence of *Rhoeo* leaf sections. Plant Physiol. **46**, 806—811.
de Vries, H., 1885: Plasmolytische Studien über die Wand der Vakuolen. Jahrb. wiss. Bot. **16**, 465—598.
Diers, L., F. Schoetz, and B. Meyer, 1973: Über die Ausbildung von Gerbstoffvakuolen bei Oenotheren. Cytobiol. **7**, 10—19.
Dingle, J. T., 1969: The extracellular secretion of lysosomal enzymes. In: Lysosomes in biology and pathology (Dingle, J. T., and H. B. Fell, eds.), Vol. 2, pp. 421—436. Amsterdam-London: North-Holland Publishing Company.
— (ed.), 1973: Lysosomes in biology and pathology, Vol. 3. Amsterdam-London: North-Holland Publishing Company; New York: American Elsevier Publishing Company, Inc.
— and H. B. Fell (eds.), 1969: Lysosomes in biology and pathology, Vol. 1 and 2. Amsterdam-London: North-Holland Publishing Company.
Doi, E., R. Hayashi, and T. Hata, 1967: Purification of yeast proteinases. II. Purification and some properties of yeast proteinase C. Agr. Biol. Chem. (Tokyo) **31**, 160—169.
Doerr, I., 1969: Feinstruktur intrazellular wachsender Cuscuta-Hyphen. Protoplasma **67**, 123—137.
— and R. Kollmann, 1969: Fine structure of mycorrhiza in *Neottia nidus avis* (L.) L. C. Rich. (Orchidaceae). Planta **89**, 372—375.
— — 1973: Strukturelle Grundlage des Parasitismus bei Orobanche. 1. Wachstum der Haustorialzellen im Wirtsgewebe. Protoplasma **80**, 245—260.
Dove, L. D., 1973: Ribonucleases in vascular plants: cellular distribution and changes during development. Phytochemistry **12**, 2561—2570.
Drucker, H., 1973: Regulation of exocellular protease in *Neurospora crassa:* role of *Neurospora* proteases in induction. J. Bacteriol. **116**, 593—599.
Dyer, T. A., and P. I. Payne, 1974: Katabolism of plant cytoplasmic ribosomes: A study of the interaction between ribosomes and ribonuclease. Planta **117**, 259—268.

Eastwood, D., and D. L. Laidman, 1971: The mobilization of macronutrient elements in the germinating wheat grain. Phytochemistry **10**, 1275—1284.
— R. J. A. Tavener, and D. L. Laidman, 1969: Induction of lipase and phytase activities in the aleurone tissue of germinating wheat grains. Biochem. J. **113**, 32 P.
Eberhart, B. M., and R. S. Beck, 1970: Localization of the β-glucosidases in *Neurospora crassa*. J. Bacteriol **101**, 408—417.
— 1973: Induction of β-glucosidases in *Neurospora crassa*. J. Bacteriol. **116**, 295—303.
Ehrlich, M. A., J. F. Schafer, and H. G. Ehrlich, 1968: Lomasomes in wheat leaves infected by *Puccinia graminis* and *P. recondita*. Can. J. Bot. **46**, 17—20.
Engleman, E. M., 1966: Ontogeny of aleurone grains in cotton embryo. Amer. J. Bot. **53**, 231—237.
English, P. D., and P. Albersheim, 1969: Host-pathogen interactions. I. A correlation between α-galactosidase secretion and virulence. Plant Physiol. **44**, 217—224.
— J. B. Jurale, and P. Albersheim, 1971: Host-pathogen interactions. II. Parameters affecting polysaccharide-degrading enzyme secretion by *Colletotrichum lindemuthianum* grown in culture. Plant Physiol. **47**, 1—6.
— A. Maglotain, K. Keegstra, and P. Albersheim, 1972: A cell wall-degrading endo-polygalacturonase secreted by *Colletotrichum lindemuthianum.* Plant Physiol. **49**, 293—298.
Evans, M. L., 1974: Evidence against the involvement of galactosidase or glucosidase in auxin- or acid-stimulated growth. Plant Physiol. **54**, 213—215.

Evins, W. H., and J. E. Varner, 1971: Hormone-controlled synthesis of endoplasmic reticulum in barley aleurone cells. Proc. nat. Acad. Sci. (US) **68**, 1631—1633.
— — 1972: Hormonal control of polysome formation in barley aleurone layers. Plant Physiol. **49**, 348—352.

Fan, D. F., and G. A. MacLachlan, 1967: Studies on the regulation of cellulase activity and growth in excised pea epicotyl sections. Can. J. Bot. **45**, 1837—1844.
Figier, J., 1968: Localisation infrastructurale de la phosphomonoestérase acide dans la stipule de *Vicia faba* L. au niveau du néctaire. Planta **83**, 60—79.
— 1969: Incorporation de glycine-^{3}H chez les glandes pétiolaires de *Mercurialis annua* L. Planta **87**, 275—289.
— 1972: Localisation infrastructurale de la phosphatase acide dans les glandes pétiolaires d'*Impatiens holstii*. Planta **108**, 215—226.
Filner, T. F., and J. E. Varner, 1967: A simple and unequivocal test for *de novo* synthesis of enzymes: density labelling of barley α-amylase with $H_2^{18}O$. Proc. nat. Acad. Sci. (US) **58**, 1520—1526.
Fineran, B. A., 1970 a: An evaluation of the form of vacuoles in thin sections and freeze-etch replicas of root tips. Protoplasma **70**, 457—478.
— 1970 b: Organization of the tonoplast frozen-etched root tips. J. Ultrastruct. Res. **35**, 574—586.
— 1971: Ultrastructure of vacuolar inclusions in root tips. Protoplasma **72**, 1—18.
— 1973: Association between endoplasmic reticulum and vacuoles in frozen-etched root tips. J. Ultrastruct. Res. **43**, 75—87.
Fisher, M. L., A. J. Anderson, and P. Albersheim, 1973: Host-pathogen interactions VI. A single plant protein efficiently inhibits endopolygalacturonases secreted by *Colletotrichum lindemuthianum* and *Aspergillus urger*. Plant Physiol. **51**, 489—491.
Flores-Carreon, A., E. Reyes, and J. Ruiz-Herrera, 1970: Inducible cell-wall-bound α-glucosidase in *Mucor rouxii*. Biochim. biophys. Acta **222**, 354—360.
Freiburg, S. R., and H. E. Clark, 1955: Changes in nitrogen fixation and proteolytic enzymes of soybean plants treated with 2,4 D. Plant Physiol. **30**, 39—46.
Frey-Wyssling, A., E. Grieshaber, and K. Mühlethaler, 1963: Origin of spherosomes in plant cells. J. Ultrastruct. Res. **8**, 506—516.
Fukuhara, H., 1967: Protein synthesis in non-growing yeast. Respiratory adaption system. Biochim. biophys. Acta **134**, 143—164.

Gahan, P. B., 1965: Histochemical evidence for the presence of lysosome-like particles in root meristem cells of *Vicia faba*. J. Exper. Bot. **16**, 350—355.
— and J. McLean, 1969: Subcellular localization and possible functions of acid β-glycerophosphatases and naphthol esterases in plant cells. Planta **89**, 126—135.
Garg, G. K., and T. K. Virupaksha, 1970: Acid protease from germinated *Sorghum*. Eur. J. Biochem. **17**, 4—12.
Garibaldi, A., and D. F. Bateman, 1971: Pectic enzymes produced by *Erwinia chrysanthemi* and their effects on plant tissue. Physiol. Plant Pathol. **1**, 25—40.
Gäumann, E., 1935: Der Stoffhaushalt der Buche im Verlauf eines Jahres. Ber. schweiz. bot. Ges. **44**, 157—334.
Gay, J. L., A. D. Greenwood, and I. B. Heath, 1971: The formation and behaviour of vacuoles (vesicles) during Oosphere development and Zoospore germination in *Saprolegnia*. J. Gen. Microbiol. **65**, 233—241.
Genevès, L., 1969: Répartition de l'activité de la phosphatase acide dans le tissu sporogène d'Hypnum rusciforme en rapport avec son évolution, a l'échelle des ultrastructures. C. R. Acad. Sci. (Paris) **269**, 1066—1069.
Gerhardt, P., and J. A. Judge, 1964: Porosity of isolated cell walls of *Saccharomyces cerevisiae* and *Bacillus megaterium*. J. Bacteriol. **87**, 945—951.
Gezelius, K., 1971: Acid phosphatase localization in myxamoebae of *Dictyostelium discoideum*. Arch. Microbiol. **75**, 327—337.
— 1972: Acid phosphatase localization during differentiation in the cellular slime mold *Dictyostelium discoideum*. Arch. Microbiol. **85**, 51—60.
Gibson, R. A., and L. G. Paleg, 1972: Lysosomal nature of hormonally induced enzymes in wheat aleurone cells. Biochem. J. **128**, 367—375.

GIFFORD, E. M., and K. D. STEWART, 1968: Inclusions of the proplastids and vacuoles in the shoot apices of *Bryophyllum* and *Kalanchoë*. Amer. J. Bot. **55**, 269—279.

GIRBARDT, M., 1969: Die Ultrastruktur der Apikalregion von Pilzhyphen. Protoplasma **67**, 413—441.

GOLDSTONE, A., and H. KOENIG, 1973: Physicochemical modifications of lysosomal hydrolases during intracellular transport. Biochem. J. **132**, 267—282.

GÓRSKA-BRYLASS, A., 1965: Hydrolases in pollen grains and pollen tubes. Acta Soc. Bot. Pol. **34**, 589—604.

GRAHAM, T. A., and B. E. S. GUNNING, 1970: The localization of legumin and vicilin in bean cotyledon cells using fluorescent antibodies. Nature **228**, 81—82.

GREEN, T. R., and C. A. RYAN, 1972: Wound-induced proteinase inhibitor in plant leaves: a possible defense mechanism against insects. Science **175**, 776—777.

GREYSON, R. I., and K. R. MITCHELL, 1969: Light and electron microscope observations of a vacuolar structure associated with the floral apex of *Nigella damascena*. Can. J. Bot. **47**, 497—601.

GRIFFITHS, D. A., 1971: Hyphal structure in *Fusarium oxysporum* (Schlecht) revealed by freeze-etching. Arch. Microbiol. **79**, 93—101.

GROVE, S. N., and C. E. BRACKER, 1970: Protoplasmic organization of hyphal tips among fungi: vesicles and Spitzenkörper. J. Bacteriol. **104**, 989—1009.

— — and D. J. MORRE, 1970: An ultrastructural basis for hyphal tip growth in *Phythium ultimum*. Amer. J. Bot. **57**, 245—266.

GRUBER, P. J., R. N. TRELEASE, W. M. BECKER, and E. H. NEWCOMB, 1970: A correlative ultrastructural and enzymatic study of cotyledonary microbodies following germination of fat-storing seeds. Planta **93**, 269—288.

GÜNTHER, TH., und W. KATTNER, 1968: Über die Funktion der sauren Hefephosphatase. Z. Naturforsch. **23 b**, 77—80.

— — und H. J. MERKER, 1967: Über das Verhalten und die Lokalisation der sauren Phosphatase von Hefezellen bei Repression und Derepression. Exp. Cell Res. **45**, 133—147.

HABERLANDT, G., 1890: Die Kleberschicht des Gras-Endosperms als Diastase ausscheidendes Drüsengewebe. Ber. Deutsch. Bot. Ges. **8**, 40—48.

HÄCKER, M., 1967: Der Abbau von Speicherprotein und die Bildung von Plastiden in den Kotyledonen des Senfkeimlings (*Sinapis alba* L.) unter dem Einfluß des Phytochroms. Planta **76**, 309—325.

HAGAR, S. S., and G. A. McINTYRE, 1972: Pectic enzymes produced by *Pseudomonas fluorescens*, an organism associated with "pink eye" disease of potato tubers. Can. J. Bot. **50**, 2479—2488.

HALL, C. B., 1963: Cellulase in tomato fruits. Nature (London) **200**, 1010—1011.

HALL, J. L., 1969: Histochemical localization of β-glycerophosphatase activity in young root tips. Ann. Bot. **33**, 399—406.

— and V. S. BUTT, 1968: Localization and kinetic properties of β-glycerophosphatase in barley roots. J. Exp. Bot. **19**, 276—287.

— and C. A. M. DAVIE, 1971: Localization of acid hydrolase activity in *Zea mays* L. root tips. Ann. Bot. **35**, 849—855.

HALPERIN, W., 1969: Ultrastructural localization of acid phosphatase in cultured cells of *Daucus carota*. Planta **88**, 91—102.

HALVORSON, H. O., 1958 a: Studies on protein and nucleic acid turnover in growing cultures of yeast. Biochim. biophys. Acta **27**, 267—276.

— 1958 b: Intracellular protein and nucleic acid turnover in resting yeast cells. Biochim. biophys. Acta **27**, 255—266.

— 1960: The induced synthesis of proteins. Advances in Enzymology **22**, 99—156.

HARRIS, N., and A. D. DODGE, 1972: The effect of paraquat on flax cotyledon leaves: Changes in fine structure. Planta **104**, 201—209.

HASUNUMA, K., 1973: Repressible extracellular nucleases in *Neurospora crassa*. Biochim. biophys. Acta **319**, 288—293.

HATA, T., R. HAYASHI, and E. DOI, 1967 a: Purification of yeast proteinases. I. Fractionation and some properties of the proteinases. Agr. Biol. Chem. (Tokyo) **31**, 150—159.

— — — 1967 b: Purification of yeast proteinases.n III. Isolation and physiochemical properties of yeast proteinase A and C. Agr. Biol. Chem. (Tokyo) **31**, 357—367.

HAYASHI, R., and T. HATA, 1972: Yeast pro-proteinase C. IV. Activation by yeast proteinase A and the activation mechanism. Agr. Biol. Chem. (Tokyo) **36**, 630—638.

— S. MOORE, and W. H. STEIN, 1973: Carboxypeptidase from yeast. J. biol. Chem. **248**, 2296—2302.

— Y. OKA, and T. HATA, 1969: Yeast pro-proteinase C. Part II, Isolation and some physico-chemical properties. Agr. Biol. Chem. (Tokyo) **33**, 196—206.

HEATH, I. B., J. L. GAY, and A. D. GREENWOOD, 1971: Cell wall formation in the Saprolegniales: Cytoplasmic vesicles underlying developing walls. J. Gen. Microbiol. **65**, 225—232.

HÉBANT, CH., 1973: Acid phosphomonoesterase activities (β-glycerophosphatase and naphthol AS-MX phosphatase) in conducting tissues of bryophytes. Protoplasma **77**, 231—241.

HEFTMANN, E., 1971: Lysosomes in tomatoes. Cytobios **3**, 129—136.

HEINIGER, U., and PH. MATILE, 1974: Protease secretion in *Neurospora crassa*. Biochem. biophys. Res. Comm. **60**, 1425—1432.

HERTH, W., W. W. FRANKE, and W. J. VAN DER WOUDE, 1972: Cytochalasin stops tip growth in plants. Naturwissenschaften **59**, 38—39.

HESS, W. M., 1969: Ultrastructure of onion roots infected with *Pyrenochaeta terrestris* a fungus parasite. Amer. J. Bot. **56**, 832—845.

— and D. J. WEBER, 1973: Ultrastructure of dormant and germinated sporangiospores of *Rhizopus arrhizus*. Protoplasma **77**, 15—33.

HEYSER, W., 1971: Phloemdifferenzierung bei *Tradescantia albiflora*. Cytobiologie **4**, 186—197.

HIRAI, M., and T. ASAHI, 1973: Membranes carrying acid hydrolases in pea seedling roots. Plant Cell Physiol. **14**, 1019—1030.

HISLOP, E. C., G. V. HOAD, and S. A. ARCHER, 1973: The involvement of ethylene in plant diseases. In: Fungal pathogenicity and the plant's response (BYRDE, R. J. W., and C. V. CUTTING, eds.). London-New York: Academic Press.

HOBDAY, S. M., D. A. THURMAN, and D. J. BARBER, 1973: Proteolytic and trypsin inhibitory activities in extracts of germinating *Pisum sativum* seeds. Phytochemistry **12**, 1041—1046.

HOBSON, G. E., 1963: Pectinesterase in normal and abnormal tomato fruit. Biochem. J. **86**, 358—365.

— 1964: Polygalacturonase in normal and abnormal tomato fruit. Biochem. J. **92**, 324—332.

HOHL, H. R., 1965: Nature and development of membrane systems in food vacuoles of cellular slime molds predatory upon bacteria. J. Bacteriol. **90**, 755—765.

— and S. T. HAMAMOTO, 1969: Ultrastructure of spore differentiation in *Dictyostelium*: the prespore vacuole. J. Ultrastruct. Res. **26**, 442—453.

HOLCOMB, G. E., A. C. HILDEBRANDT, and R. F. EVERT, 1967: Staining and acid phosphatase reactions of spherosomes in plant tissue culture cells. Amer. J. Bot. **54**, 1204—1209.

HOLZER, H., T. KATSUNUMA, E. G. SCHÖTT, A. R. FERGUSON, A. HASILIK, and H. BETZ, 1973: Studies on a tryptophan synthase inactivating system from yeast. Advan. Enz. Regul. **11**, 53—60.

HORTON, R. F., and D. J. OSBORNE, 1967: Senescence, abscission, and cellulase activity in *Phaseolus vulgaris*. Nature **214**, 1086—1088.

HOTTA, Y., A. BASSEL, and H. STERN, 1965: Nuclear DNA and cytoplasmic DNA from tissues of higher plants. J. Cell Biol. **27**, 451—457.

HUFFAKER, R. C., and L. W. PETERSON, 1974: Protein turnover in plants and possible means of its regulation. Ann. Rev. Plant Physiol. **25**, 363—392.

HULME, M. A., and D. W. STRANKS, 1971: Regulation of cellulase production by *Myrothecium verrucaria* grown on non-cellulosic substrates. J. Gen. Microbiol. **69**, 145—156.

HUMBERT, CL., et M. GUYOT, 1972: Modifications ultrastructurales des cellules stomatiques d'*Anemia rotundifolia* Schrad. C. R. Acad. Sci. (Paris) **274**, 380—382.

ITEN, W., 1969: Zur Funktion hydrolytischer Enzyme bei der Autolyse von *Coprinus*. Ber. schweiz. bot. Ges. **79**, 175—198.
— and PH. MATILE, 1970: Role of chitinase and other lysosomal enzymes of *Coprinus lagopus* in the autolysis of fruiting bodies. J. Gen. Microbiol. **61**, 301—309.

JACKS, T. J., L. Y. YATSU, and A. M. ALTSCHUL, 1967: Isolation and characterization of peanut spherosomes. Plant Physiol. **42**, 585—597.
JACKSON, M. B., and D. J. OSBORNE, 1972: Abscisic acid, auxin, and ethylene in explant abscission. J. Exp. Bot. **23**, 849—862.
JACOBS, M. M., 1973: Enzymatic properties of a repressible acid phosphatase in *Neurospora crassa*. Texas Rep. Biol. Med. **31**, 5—17.
— J. F. NYC, and D. M. BROWN, 1971: Isolation and chemical properties of a repressible acid phosphatase in *Neurospora crassa*. J. biol. Chem. **246**, 1419—1425.
JACOBSEN, J. V., 1973: Interactions between gibberellic acid, ethylene, and abscisic acid in control of amylase synthesis in barley aleurone layers. Plant Physiol. **51**, 198—202.
— and J. E. VARNER, 1967: Gibberellic acid-induced synthesis of protease by isolated aleurone layers of barley. Plant Physiol. **42**, 1596—1600.
JACQUES, P. J., 1969: Lysosomes and homeostatic regulation. In: Ciba foundation symposium on homeostatic regulators, pp. 180—193 (WOLSTENHOLME, G. E. W., and J. KNIGHT, eds.). London: J. & A. Churchill, Ltd.
JENSEN, W. A., 1968: Cotton embryogenesis: the zygote. Planta **79**, 346—366.
JOHNSON, B. F., 1968: Lysis of yeast cell walls induced by 2-deoxyglucose and their sites of glucan synthesis. J. Bacteriol. **95**, 1169—1172.
JOHNSON, K. D., D. DANIELS, M. J. DOWLER, and D. L. RAYLE, 1974: Activation of *Avena* coleoptile cell wall glycosidases by hydrogen ions and auxin. Plant Physiol. **53**, 224—228.
— and H. KENDE, 1971: Hormonal control of lecithin synthesis in barley aleurone cells: Regulation of the CDP-choline pathway by gibberellin. Proc. nat. Acad. Sci. (US) **68**, 2674—2677.
JOHNSON, L. B., B. L. BRANNAMAN, and F. P. ZSCHEILE, 1968: Protein and enzyme changes in wheat leaves following infection with *Puccinia recondita*. Phytopathology **58**, 578—583.
JONES, R. L., 1969 a: The effect of ultracentrifugation on fine structure and α-amylase production in barley aleurone cells. Plant Physiol. **44**, 1428—1438.
— 1969 b: Gibberellic acid and the fine structure of barley aleurone cells. I. Changes during the lag-phase of α-amylase synthesis. Planta **87**, 119—133.
— 1971: Gibberellic acid-enhanced release of β-1,3-glucanase from barley aleurone cells. Plant Physiol. **47**, 412—416.
— 1972: Fractionation of the enzyme of the barley aleurone layer: Evidence for a soluble mode of enzyme release. Planta **103**, 95—109.
— and J. E. ARMSTRONG, 1971: Evidence for osmotic regulation of hydrolytic enzyme production in germinating barley seeds. Plant Physiol. **48**, 137—142.
— and J. M. PRICE, 1970: Gibberellic acid and the fine structure of barley aleurone cells. III. Vacuolation of aleurone cell during the phase of ribonuclease release. Planta **94**, 191—202.
JONES, T. M., A. J. ANDERSON, and P. ALBERSHEIM, 1972: Host-pathogen interactions. IV. Studies on the polysaccharide-degrading enzymes secreted by *Fusarium oxysporum* f. sp. *lycopersici*. Physiol. Plant Pathol. **2**, 153—166.

KARR, A. L., JR., and P. ALBERSHEIM, 1970: Polysaccharide-degrading enzymes are unable to attack plant cell walls without prior action by a "wall-modifying enzyme". Plant Physiol. **46**, 69—80.
KATZ, M., and L. ORDIN, 1967 a: Metabolic turnover in cell wall constituents of *Avena sativa* L. coleoptile sections. Biochim. biophys. Acta **141**, 118—125.
— — 1967 b: A cell wall polysaccharide-hydrolyzing enzyme system in *Avena sativa* L. coleoptiles. Biochim. biophys. Acta **141**, 126—134.

Keegstra, K., and P. Albersheim, 1970: The involvement of glucosidases in the cell wall metabolism of suspension-cultured *Acer pseudoplatanus* cells. Plant Physiol. **45**, 675—678.
— P. D. English, and P. Albersheim, 1972: Four glycosidases secreted by *Collectotrichum lindemuthianum*. Phytochemistry **11**, 1873—1880.
Kelker, H. Ch., and Ph. Filner, 1971: Regulation of nitrite reductase and its relationship to the regulation of nitrate reductase in cultured tobacco cells. Biochim. biophys. Acta **252**, 69—82.
Kende, H., 1971: The cytokinins. Int. Rev. Cytol. **31**, 301—338.
— and B. Baumgartner, 1974: Regulation of ageing in flowers of *Ipomoea tricolor* by ethylene. Planta **116**, 279—289.
Kern, H., und S. Naef-Roth, 1971: Phytolysin, ein durch pflanzenpathogene Pilze gebildeter mazerierender Faktor. Phytopath. Z. **71**, 231—246.
Keusch, L., 1968: Die Mobilisierung des Reservemannans im keimenden Dattelsamen. Planta **78**, 321—350.
Khoo, U., and M. J. Wolf, 1970: Origin and development of protein granules in maize endosperm. Amer. J. Bot. **57**, 1042—1050.
Kidby, D. K., and R. Davies, 1970: Invertase and disulphide bridges in the yeast wall. J. gen. Microbiol. **61**, 327—333.
Kirsi, M., and J. Mikola, 1971: Occurrence of proteolytic inhibitors in various tissues of barley. Planta **96**, 281—291.
Kivilaan, A., R. S. Bandurski, and A. Schulze, 1971: A partial characterization of an autolytically solubilized cell wall glucan. Plant Physiol. **48**, 389—393.
— T. C. Beaman, and R. S. Bandurski, 1961: Enzymatic activities associated with cell wall preparations from corn coleoptiles. Plant Physiol. **36**, 605—610.
Klapper, B. F., D. M. Jameson, and R. M. Mayer, 1973: Factors affecting the synthesis and release of the extracellular protease of *Aspergillus oryzae* NRRL 2160. Biochim. biophys. Acta **304**, 513—519.
Klis, F. M., 1971: β-glucosidase activity located at the cell surface in callus of *Convolvulus arvensis*. Physiol. Plant. **25**, 253—257.
Knee, M., 1973: Polysaccharide changes in cell walls of ripening apples. Phytochemistry **12**, 1543—1550.
Knox, R. B., and J. Heslop-Harrison, 1970: Pollen-wall proteins: localization of enzymic activity. J. Cell. Sci. **6**, 1—27.
— — 1971: Pollen-wall proteins: electron-microscopic localization of acid phosphatase in the intine of *Crocus vernus*. J. Cell Sci. **8**, 727—733.
Koehler, D. E., and J. E. Varner, 1973: Hormonal control of orthophosphate incorporation into phospholipids of barley aleurone layers. Plant Physiol. **52**, 208—214.
Koenig, H., 1969: Lysosomes in the nervous system. In: Lysosomes in biology and pathology (Dingle, J. T., and H. B. Fell, eds.), Vol. 2, 111—162. Amsterdam-London: North-Holland Publ. Comp.
Kuo, M. H., and H. J. Blumenthal, 1961: Purification and properties of an acid phosphomonoesterase from *Neurospora crassa*. Biochim. biophys. Acta **52**, 13—29.

Lampen, J. O., 1968: External enzymes of yeast: their nature and formation. Anton Leeuwenhoek J. Microbiol. **34**, 1—18.
— S.-C. Kuo, F. R. Cano, and J. S. Tkacz, 1972: Structural features in synthesis of external enzymes by yeast. Proc. 4th Int. Fermentation Symposium, Kyoto, pp. 122—127.
Lee, S., A. Kivilaan, and R. S. Bandurski, 1967: *In vitro* autolysis of plant cell walls. Plant Physiol. **42**, 968—972.
Lenney, J. F., 1956: A study of two yeast proteinases. J. biol. Chem. **221**, 919—930.
— and J. M. Dalbec, 1967: Purification and properties of two proteinases from *Saccharomyces cerevisiae*. Arch. Biochem. Biophys. **120**, 42—48.
— — 1969: Yeast proteinase B: Identification of the inactive form as an enzyme inhibitor complex. Arch. Biochem. Biophys. **129**, 407—409.
— Ph. Matile, A. Wiemken, M. Schellenberg, and J. Meyer, 1974: Activities and cellular localization of yeast proteases and their inhibitors. Biochem. biophys. Res. Comm. **60**, 1378—1383.

LEWIS, L. N., and J. E. VARNER, 1970: Synthesis of cellulase during abscission of *Phaseolus vulgaris* leaf explants. Plant Physiol. **46**, 194—199.

LITTLEFIELD, L. J., and CH. E. BRACKER, 1970: Continuity of host plasma membrane around haustoria of *Melampsora lini*. Mycologia **62**, 609—614.

LOESCHER, W., and D. J. NEVINS, 1972: Auxin-induced changes in *Avena* coleoptile cell wall composition. Plant Physiol. **50**, 556—563.

LOTT, J. N. A., and C. M. VOLLMER, 1973: Changes in the cotyledons of *Cucurbita maxima* during germination. IV. Protein bodies. Protoplasma **78**, 255—272.

LOWRY, R. J., and A. S. SUSSMAN, 1968: Ultrastructural changes during germination of ascospores of *Neurospora tetrasperma*. J. gen. Microbiol. **51**, 403—409.

— T. L. DURKEE, and A. S. SUSSMAN, 1967: Ultrastructural studies of microconidium formation in *Neurospora crassa*. J. Bacteriol. **94**, 1757—1763.

LUI, N. S. T., and A. M. ALTSCHUL, 1967: Isolation of globoids from cottonseed aleurone grain. Arch. Biochem. Biophys. **121**, 678—684.

LÜSCHER, A., and PH. MATILE, 1974: Studies on the (vacuolar?) localization of RNase and other enzymes in *Acetabularia*. Planta **118**, 323—332.

MACLACHLAN, G. A., and M. YOUNG, 1962: Breakdown and synthesis of cell walls during growth. Nature **195**, 1319—1321.

MADDOX, I. S., and J. S. HOUGH, 1971: Yeast glucanase and mannanase. J. Inst. Brew. **77**, 44—47.

MAIER, K., and U. MAIER, 1972: Localization of beta-glycerophosphatase and Mg++-activated adenosine triphosphatase in moss haustorium, and the relation of these enzymes to the wall labyrinth. Protoplasma **75**, 91—112.

MAITRA, S. C., and N. DE DEEPESH, 1972: Ultrastructure of root cap cells: formation and utilization of lipid. Cytobios **5**, 111—118.

MALIK, C. P., P. P. SOOD, and H. B. TEWARI, 1969: Occurrence of lysosome-like bodies in plant cells-acid phosphatase reaction. Z. Biol. **116**, 264—268.

MALKOFF, D. B., and D. E. BUETOW, 1964: Ultrastructural changes during carbon starvation in *Euglena gracilis*. Exp. Cell Res. **35**, 58—68.

MANOCHA, M. S., and J. R. COLVIN, 1967: Structure and composition of the cell walls of *Neurospora crassa*. J. Bacteriol. **94**, 202—212.

MARINOS, N. G., 1963: Vacuolation in plant cells. J. Ultrastruct. Res. **9**, 177—185.

MARTIN, C., and K. V. THIMANN, 1972: The role of protein synthesis in the senescence of leaves. I. The formation of protease. Plant Physiol. **49**, 64—71.

MARTY, F., 1970: Rôle du système membranaire vacuolaire dans la differenciation des lacticifères d'*Euphorbia characias* L. C. R. Acad. Sci. (Paris) **271**, 2301—2304.

— 1971 a: Différenciation des plastes dans les laticiferes d'*Euphorbia characias* L. C. R. Acad. Sci. (Paris) **272**, 223—226.

— 1971 b: Vésicules autophagiques des laticifères différenciés d'*Euphorbia characias* L. C. R. Acad. Sci. (Paris) **272**, 399—402.

— 1971 c: Peroxisomes et compartiment lysosomal dans les cellules du méristème radiculaire d'*Euphorbia characias* L.: une étude cytochimique. C. R. Acad. Sci. (Paris) **273**, 2504—2507.

— 1972: Distributions des activités phosphatasiques acides au cours du processus d'autophagie cellulaire dans les cellules du méristème radiculaire d'*Euphorbia characias* L. C. R. Acad. Sci. (Paris) **274**, 206—209.

— 1973: Mise en évidence d'un appareil provacuolaire et de son rôle dans l'autophagie cellulaire et l'origine des vacuoles. C. R. Acad. Sci. (Paris) **276**, 1549—1552.

MARZLUF, G. A., 1972: Control of the synthesis, activity and turnover of enzymes of sulfur metabolism in *Neurospora crassa*. Arch. Biochem. Biophys. **150**, 714—724.

MASUDA, Y., S. OI, and Y. SATOMURA, 1970: Further studies on the role of cell-wall-degrading enzymes in cell-wall loosening in oat coleoptiles. Plant Cell Physiol. **11**, 631—638.

MATCHETT, W. H., and J. F. NANCE, 1962: Cell wall breakdown and growth in pea seedling stems. Amer. J. Bot. **49**, 311—319.

MATHESON, N. K., and S. STROTHER, 1969: The utilization of phytate by germinating wheat. Phytochemistry **8**, 1349—1356.

MATILE, PH., 1965: Intrazelluläre Lokalisation proteolytischer Enzyme von *Neurospora crassa*. I. Funktion und subzelluläre Verteilung proteolytischer Enzyme. Z. Zellforsch. Mikroskop. Anat. **65**, 884—896.

— 1966 a: Inositol defiency resulting in death: an explanation of its occurrence in *Neurospora crassa*. Science **151**, 86—88.

— 1966 b: Enzyme der Vakuolen aus Wurzelzellen von Maiskeimlingen. Ein Beitrag zur funktionellen Bedeutung der Vakuole bei der intrazellulären Verdauung. Z. Naturforsch. Sect. **B 21**, 871—878.

— 1968 a: Lysosomes of root tip cells in corn seedlings. Planta **79**, 181—196.

— 1968 b: Aleurone vacuoles as lysosomes. Z. Pflanzenphysiol. **58**, 365—368.

— 1969: Utilization of peptides in yeasts. In: Yeasts, Proc. 2nd Symposium on yeasts, Bratislava 1966, pp. 503—508. Bratislava: Publishing House of the Slovak Academy of Sciences.

— 1970: Recent progress in the study of yeast cytology. Proc. 2nd int. Symposium yeast protoplasts, Brno. Acta Facultatis Medicae Universitatis Bruniensis **37**, 17—23.

— 1971: Vacuoles, lysosomes of *Neurospora*. Cytobiologie **3**, 324—330.

— 1973: Regulation und Bedeutung der Sekretion von Hydrolasen. Ber. deutsch. bot. Ges. **86**, 241—255.

— 1974: Cell wall degradation in senescing tobacco leaf discs. Experientia **30**, 98—99.

— J. P. BALZ, E. SEMADENI, and M. JOST, 1965: Isolation of spherosomes with lysosome characteristics from seedlings. Z. Naturforsch. Sect. **B 20**, 693—698.

— B. JANS, and R. RICKENBACHER, 1970: Vacuoles of *Chelidonium latex*: lysosomal property and accumulation of alcaloids. Biochem. Physiol. Pflanzen **161**, 447—458.

— and H. MOOR, 1968: Vacuolation: origin and development of the lysosomal apparatus in root tip cells. Planta **80**, 159—175.

— — and C. F. ROBINOW, 1969: Yeast cytology. In: The yeasts, pp. 219—302 (ROSE A. H., and J. S. HARRISON, eds.), Vol. 1. London-New York: Academic Press.

— and J. SPICHIGER, 1968: Lysosomal enzymes in spherosomes (oil droplets) of tobacco endosperm. Z. Pflanzenphysiol. **58**, 277—280.

— and A. WIEMKEN, 1967: The vacuole as the lysosome of the yeast cell. Arch. Mikrobiol. **56**, 148—155.

— — and W. GUYER, 1971: A lysosomal aminopeptidase isozyme in differentiating yeast cells and protoplasts. Planta **96**, 43—53.

— and F. WINKENBACH, 1971: Function of lysosomes and lysosomal enzymes in the senescing corolla of the morning glory (*Ipomoea purpurea*). J. Exp. Bot. **22**, 759—771.

MAYER, A. M., and Y. SHAIN, 1968: Zymogen granules in enzyme liberation and activation in pea seeds. Science **162**, 1283—1284.

McCULLY, E. K., and C. E. BRACKER, 1972: Apical vesicles in growing bud cells of heterobasidiomycetous yeasts. J. Bacteriol. **109**, 922—926.

McLEAN, J., and P. B. GAHAN, 1970: The distribution of caid phosphatases and esterases in differentiating roots of *Vicia faba*. Histochemie **24**, 41—49.

MESQUITA, J. F., 1969: Electron microscope study of the origin and development of the vacuoles in root-tip cells of *Lupinus albus* L. J. Ultrastruct. Res. **26**, 242—250.

— 1972: Ultrastructure de formations comparables aux vacuoles autophagiques dans les cellules des racines de l'*Allium cepa* L. et du *Lupinus albus* L. Cytologia (Tokyo) **37**, 95—110.

MEYER, H., A. M. MAYER, and E. HAREL, 1971: Acid phosphatases in germinating lettuce— Evidence for partial activation. Physiol. Plant. **24**, 95—101.

MEYER, J., and PH. MATILE, 1974 a: Subcellular distribution of yeast invertase isoenzymes. Arch. Microbiol. **103**, 51—55.

— — 1974 b: Regulation of isoenzymes and secretion of invertase in baker's yeast. Biochem. Physiol. Pflanzen **166**, 377—385.

MITTELHEUSER, C. J., and R. F. M. VAN STEVENINCK, 1971: The ultrastructure of wheat leaves. I. Changes due to natural senescence and the effects of kinetin and ABA on detached leaves incubated in the dark. Protoplasma **73**, 239—252.

— — 1972: Ultrastructural changes in naturally senescing leaves compared with changes induced by (RS)-abscisic acid. In: Plant growth substances (CARR, D. J. ed.). New York: Plenum.

MOLLENHAUER, H. H., and J. D. MORRÉ, 1966: Golgi apparatus and plant secretion. Ann. Rev. Plant Physiol. **17**, 27—46.

— and C. TOTTEN, 1970: Studies on seeds. V. Microbodies, glyoxysomes, and ricinosomes of castor bean endosperm. Plant Physiol. **46**, 794—799.

— — 1971 a: Studies on seeds. I. Fixation of seeds. J. Cell Biol. **48**, 387—394.

— — 1971 b: Studies on seeds. II. Origin and degradation of lipid vesicles in pea and bean cotyledons. J. Cell Biol. **48**, 395—405.

MOOR, H., 1967: Endoplasmic reticulum as the initiator of bud formation in yeast (*S. cerevisiae*). Arch. Mikrobiol. **57**, 135—146.

MOORE, A. E., and B. A. STONE, 1972: Effect of senescence and hormone treatment on the activity of a β-1,3-glucan hydrolase in *Nicotiana glutinosa* leaves. Planta **104**, 93—109.

MORRÉ, J. D., H. H. MOLLENHAUER, and C. B. BRACKER, 1971: Origin and continuity of Golgy apparatus. In: Results and problems in cell differentiation, pp. 82—126 (Origin and continuity of cell organelles, Vol. 2). (REINERT, J., and H. URSPRUNG, eds.) Berlin-Heidelberg-New York: Springer.

MORRIS, G. F. I., D. A. THURMAN, and D. BOULTER, 1970: The extraction and chemical composition of aleurone grains (protein bodies) isolated from seeds of *Vicia faba*. Phytochemistry **9**, 1707—1714.

MORTON, R. K., and J. K. RAISON, 1963: A complete intracellular unit for incorporation of amino acids into storage protein utilizing adenosine triphosphate generated from phylate. Nature **200**, 429—433.

MOTHES, K., 1960: Über das Altern der Blätter und die Möglichkeit ihrer Wiederverjüngung. Naturwissenschaften **47**, 337—351.

MUKHERJI, S., B. DEY, A. K. PAUL, and S. M. SIRKAR, 1971: Changes in phosphorus fractions and phytase activity of rice seeds during germination. Physiol. Plant **25**, 94—97.

MÜNTZ, K., and G. SCHOLZ, 1974: Speicherproteine und Proteinspeicherung in pflanzlichen Samen. Biol. Rundschau **12**, 225—244.

MUSE, R. R., H. B. COUCH, L. D. MOORE, and B. D. MUSE, 1972: Pectolytic and cellulolytic enzymes associated with *Helminthosporium* leaf spot on Kentucky bluegrass. Can. J. Microbiol. **18**, 1091—1098.

MUTO, S., and H. BEEVERS, 1974: Lipase activities in castor bean endosperm during germination. Plant Physiol. **54**, 23—28.

NAKANO, M., and T. ASAHI, 1972: Subcellular distribution of hydrolase in germinating cotyledons. Plant Cell. Physiol. **13**, 101—110.

NEHEMIAH, J. L., 1973: Localization of acid phosphatase activity in the basidia of *Coprinus micaceus*. J. Bacteriol. **115**, 443—446.

NEUMANN, N. P., and J. O. LAMPEN, 1969: The glycoprotein structure of yeast invertase. Biochemistry (USA) **8**, 3552—3556.

NEVINS, D. J., 1970: Relation of glycosidases to bean hypocotyl growth. Plant Physiol. **46**, 458—462.

— P. D. ENGLISH, and P. ALBERSHEIM, 1968: Changes in cell wall polysaccharides associated with growth. Plant Physiol. **43**, 914—922.

NOLAN, R. A., and A. K. BAL, 1974: Cellulase localization in hyphae of *Achlya ambisexualis*. J. Bacteriol. **117**, 840—843.

NORGAARD, M. J., and M. W. MONTGOMERY, 1968: Some esterases of the pea (*Pisum sativum* L.). Biochim. biophys. Acta **151**, 587—596.

NYC, J. F., R. J. KADNER, and B. J. CROCKEN, 1966: A repressible alkaline phosphatase in *Neurospora crassa*. J. biol. Chem. **241**, 1468—1472.

O'BRIEN, T. P., 1969: Further observations on hydrolysis of the cell wall in the Xylem. Protoplasma **69**, 1—14.

— and K. V. THIMANN, 1967: Observations on the fine structure of the oat coleoptile. III. Correlated light and electron microscopy of the vascular tissues. Protoplasma **63**, 443—478.

O'DAY, D. H., 1973: Intracellular localization and extracellular release of certain lysosomal enzyme activities from *Amoebae* of the cellular slime mold *Polysphondylium pallidum*. Cytobios. **7**, 223—232.

OLIVER, P. T. P., 1973: Influence of cytochalasin B on hyphal morphogenesis of *Aspergillus nidulans*. Protoplasma **76**, 279—281.

ONOFEGHARA, F. A., 1973: Histochemical localization of enzymes in *Tapinanthus Bangwensis* and its hosts: Alkaline phosphatase and β-glucosidase. Bot. Gaz. **134**, 39—46.

ÖPIK, H., 1968: Development of cotyledon cell structure in kipening *Phaseolus vulgaris* seeds. J. Exp. Bot. **19**, 64—76.

ORY, R. L., 1969: Acid lipase of the castor bean. Lipids **4**, 177—185.

— and K. W. HENNINGSEN, 1969: Enzymes associated with protein bodies isolated from ungerminated barley seeds. Plant Physiol. **44**, 1488—1498.

— L. Y. YATSU, and H. W. KIRCHNER, 1968: Association of lipase activity with the spherosomes of *Ricinus communis*. Arch. Biochem. Biophys. **123**, 255—264.

PALADE, G. E., P. SIEKEVITZ, and L. G. CARO, 1962: Structure, chemistry and function of the pancreas exocrine cell. Ciba foundation Symp. on the exocrine pancreas, pp. 23—49.

PAYNE, P. I., and D. BOULTER, 1974: Katabolism of plant cytoplasmic ribosomes: RNA breakdown in senescent cotyledons of germinating broad-bean seedlings. Planta **117**, 251—258.

PICKETT-HEAPS, J. D., 1967 a: Further observations on the Golgi apparatus and its functions in cells of the wheat seedling. J. Ultrastruct. Res. **18**, 287—303.

— 1967 b: Ultrastructure and differentiation in *Chara* sp. I. Vegetative cells. Austral. J. Biol. Sci. **20**, 539—551.

PILET, P. E., 1970: The effect of auxin and abscisic acid on the catabolism of RNA. J. Exp. Bot. **21**, 446—451.

PITT, D., 1968: Histochemical demonstration of certain hydrolytic enzymes within cytoplasmic particles of *Botrytis cinerea* Fr. J. gen. Microbiol. **52**, 67—75.

— 1971: Purification of a ribonuclease from potato tubers and its use as an antigen in the immunochemical assay of its protein following tuber damage. Planta **101**, 333—351.

— 1973: Solubilization of molecular forms of lysosomal acid phosphatase of *Solanum tuberosum* L. leaves during infection by *Phytophtora infestans* (Mont.) de Bary. J. gen. Microbiol. **77**, 117—126.

— 1974: Activation and *de novo* synthesis of ribonuclease following mechanical damage to leaves of *Solanum tuberosum* L. Planta **117**, 43—56.

— and C. COOMBES, 1968: The disruption of lysosome-like particles of *Solanum tuberosum* cells during infection by *Phytophtora erythroseptica Pethybr*. J. gen. Microbiol. **53**, 197—204.

— — 1969: Release of hydrolytic enzymes from cytoplasmic particles of *Solanum* tuber tissue during infection by tuberrotting fungi. J. gen. Microbiol. **56**, 321—329.

— and M. GALPIN, 1971: Increase in ribonuclease activity following mechanical damage to leaf and tuber tissues of *Solanum tuberosum* L. Planta **101**, 317—332.

— and M. GALPIN, 1973: Isolation and properties of lysosomes from dark-grown potato shoots. Planta **109**, 233—258.

— — 1973: Role of lysosomal enzymes in pathogenicity. Proc. third long Ashton Symposium, pp. 449—467.

— and P. J. WALKER, 1967: Particulate localization of acid phosphatase in fungi. Nature **215**, 783—784.

PLADYS, D., et TH. ESQUERRE-TUGAYE, 1973: Activité protéolytique de *Colletotrichum lagenarium*: mise en évidence et évolution dans le mycélium et le filtrat de culture d'une souche pathogène. C. R. Acad. Sci. (Paris) **277**, 2357—2360.

POLANY, M., 1968: Life's irreducible structure. Science **160**, 1308—1312.

POLITIS, D. J., and H. WHEELER, 1973: Ultrastructural study of penetration of maize leaves by *Colletotrichum graminicola*. Physiol. Plant Pathol. **3**, 465—471.

PORTER, K. R., and R. D. MACHADO, 1960: Form and distribution of ER during mitosis in cells of onion root tip. J. biophys. biochem. Cytol. **7**, 167—180.

POUX, N., 1962: Nouvelles observations sur la nature et l'origine de la membrane vacuolaire des cellules végétales. J. de Microscopie (Paris) **1**, 55—66.

— 1963 a: Localisation de la phosphatase acide dans les cellules meristematiques de blé (*Triticum vulgare* Vill.). J. Microscopie (Paris) **2**, 485—489.

Poux, N., 1963 b: Localisation des phosphates et de la phosphatase acide dans les cellules des embryons de blé (*Triticum vulgare* Vill.) lors de la germination. J. Microscopie (Paris) **2**, 557—568.
— 1963 c: Sur la présence d'enclaves cytoplasmiques in voie de dégénerescence dans les vacuoles des cellules végétales. C. R. Acad. Sci. (Paris) **257**, 736—738.
— 1965: Localisation de l'activité phosphatasique acide et des phosphates dans les grains d'Aleurone. I. Grains d'Al. renferment à la fois globoïdes et cristalloïdes. J. Microscopie (Paris) **4**, 771—782.
— 1966: Ultrastructure localisation of aryl sulfatase activity in plant meristematic cells. J. Histochem. Cytochem. **14**, 932—933.
— 1969: Localisation d'activités enzymatiques dans les cellules du méristème radiculaire de *Cucumis sativus* L. II. Activité peroxydasique. J. Microscopie (Paris) **8**, 855—866.
— 1970: Localisation d'activités enzymatiques dans le méristème radiculaire de *Cucumis sativus* L. III. Activité phosphatasique acide. J. Microscopie (Paris) **9**, 407—434.
Pujarniscle, S., 1968: Caractère lysosomal des lutoides du latex d'*Hevea brasiliensis,* Mül. Arg. Physiol. Végét. **6**, 27—46.
Pusztai, A., and I. Duncan, 1971: Changes in proteolytic enzyme activities and transformation of nitrogenous compounds in the germinating seeds of kidney bean (*Phaseolus vulgaris*). Planta **96**, 317—325.

Racusen, D., and M. Foote, 1970: An endopeptidase of bean leaves. Can. J. Bot. **48**, 1017—1021.
Radley, M., 1969: The effect of the endosperm on the formation of the gibberellin by barley embryos. Planta **86**, 218—223.
Ragetli, H. W. J., M. Weintraub, and E. Lo, 1970: Degeneration of leaf cells resulting from starvation after excision. I. Electron microscopic observations. Can. J. Bot. **48**, 1913—1922.
Rasmussen, G. K., 1973: Changes in cellulase and pectinase activities in fruit tissues and separation zones of citrus treated with cycloheximide. Plant Physiol. **51**, 626—628.
Ray, P. M., T. L. Shininger, and M. M. Ray, 1969: Isolation of β-glucan synthetase particles from plant cells and identification with Golgi membranes. Nat. Acad. Sci. (USA) **64**, 605—612.
Rayle, D. L., 1973: Auxin-induced hydrogen ion secretion in *Avena* coleoptiles and its implications. Planta **114**, 63—73.
— and R. Cleland, 1970: Enhancement of wall loosening and elongation by acid solutions. Plant. Physiol. **46**, 250—253.
Reddi, K. K., 1966: Ribonuclease induction in cells transformed by *Agrobacterium fumefaciens.* Proc. nat. Acad. Sci. (USA) **56**, 1207—1214.
Reid, J. S. G., 1971: Reserve carbohydrate metabolism in germinating seeds of *Trigonella foenum-graecum* L. (Leguminosae). Planta **100**, 131—142.
— and H. Meier, 1972: The function of the aleurone layer during galactomannan mobilization in germinating seeds of fenugreek (*Trigonella foenum-graecum* L.), crimson clover (*Trifolium incarnatum* L.), and lucerne (*Medicago sativa* L.): a correlation biochemical and ultrastructural study. Planta **106**, 44—60.
— — 1973 a: Formation of the endosperm galactomannan in leguminous seeds: preliminary communications. Caryologia **25**, 219—222.
— — 1973 b: Enzymic activities and galactomannan mobilization in germinating seeds of fenugreek (*Trigonella foenum-graecum* L. Leguminosae). Secretion of α-galactosidase and β-mannosidase by the aleurone layer. Planta **112**, 301—308.
Reid, M. S., and R. L. Bieleski, 1970: Changes in phosphatase activity in phosphorous-deficient *Spirodela.* Planta **94**, 273—281.
Reiss, J., 1969: Cytochemischer Nachweis von Hydrolasen in Pilzzellen. I. Glycosidasen. Histochemie **18**, 12—23.
— 1971: Lysosome-like particles in *Geotrichum candidum:* A cytochemical study. Z. Allg. Mikrobiol. **11**, 319—323.
— 1972: Cytochemischer Nachweis von Hydrolasen in Pilzzellen. II. Aminopeptidase. Acta Histochem. **39**, 277—285.

REISS, J., 1974: Cytochemical detection of hydrolases in fungus cells. III. Arylsulfatase. J. Histochem. Cytochem. **22**, 183—188.

RIDGE, E. H., and A. D. ROVIRA, 1971: Phosphatase activity of intact young wheat roots under sterile and nonsterile conditions. New Phytol. **70**, 1017—1026.

RIES, S. M., and P. ALBERSHEIM, 1973: Purification of a protease secreted by *Colletotrichum lindemuthianum*. Phytopathology **63**, 625—629.

RIOV, J., 1974: A polygalacturonase from citrus leaf explants. Plant Physiol. **53**, 312—316.

ROBERTS, D. W. A., 1970: A survey of the wheat leaf phosphatases using gel filtration with sephadex G 200. Enzymologia **39**, 151—165.

ROCK, G. D., and B. F. JOHNSON, 1970: Activity and location of two enzyme fractions during the culture cycle of *Schizosaccharomyces pombe*. Can. J. Microbiol. **16**, 187—191.

ROGGEN, H. P. J. R., and R. G. STANLEY, 1969: Cell-wall-hydrolysing enzymes in wall formation as measured by pollen-tube extension. Planta **84**, 295—303.

ROHRINGER, R., D. J. SAMBORSKI, and C. O. PERSON, 1961: Ribonuclease activity in rusted wheat leaves. Can. J. Bot. **39**, 775—784.

RUESINK, A. W., 1969: Polysaccharidases and the control of cell wall elongation. Planta **89**, 95—107.

RUINEN, J., H. DEINEMA, and CH. VAN DER SCHEER, 1968: Cellular and extracellular structures in *Cryptococcus laurentii* and *Rhodotorula species*. Can. J. Microbiol. **14**, 1133—1137.

RYAN, C. A., 1973: Proteolytic enzymes and their inhibitors in plants. Ann. Rev. Plant Physiol. **24**, 173—196.

— and O. C. HUISMAN, 1967: Chymotrypsin inhibitor I from potatoes: a transient protein component in leaves of young potato plants. Nature **214**, 1047—1049.

— and L. K. SHUMWAY, 1971: Studies on the structure and function of *Chymotrypsin* inhibitor I in the *Solanaceae* family. Proc. Internat. Res. Conf. on Proteinase Inhibitors, Munich, Nov. 1970, pp. 175—188. (FRITZ, H., and H. TSCHESCHER eds.) Berlin-New York: Walter de Gruyter.

SACHER, J. A., 1967: Studies of permeability RNA and protein turnover during ageing of fruit and leaf tissues. Symp. Soc. exp. Biol. **21**, 269—303.

— 1969: Hormonal control of senescence of bean endocarp: suppression of RNase. Plant Physiol. **44**, 313—314.

— 1973: Senescence and postharvest physiology. Ann. Rev. Plant Physiol. **24**, 197—224.

— and D. D. DAVIES, 1974: Demonstration of de novo synthesis of RNase in *Rhoeo* leaf sections by deuterium oxide labelling. Plant Cell Physiol. **15**, 157—161.

— and S. O. SALMINEN, 1969: Comparative studies of effect of auxin and ethylene on permeability and synthesis of RNA and protein. Plant Physiol. **44**, 1371—1377.

SAHULKA, J., and K. BENES, 1969: Fractions of non-specific esterase in root tip of *Vicia faba* L. revealed by disc electrophoresis in acrylamide gel. Biologia Plantarum (Praha) **11** (1), 23—33.

SAMPSON, M., and D. D. DAVIES, 1966: Synthesis of a metabolically labile DNA in the maturing root cells of *Vicia faba*. Exp. Cell Res. **43**, 669—673.

SARKAR, S. K., and R. K. PODDAR, 1965: Non-conservation of H^3-thymidine label in the DNA of growing yeast cells. Nature **207**, 550—551.

SASSEN, M. M. A., 1965: Breakdown of the plant cell wall during the cell-fusion process. Acta Botanica Neerl. **14**, 165—196.

SCALA, J., D. W. SCHWAB, and F. E. SEMERSKY, 1968: The fine structure of the digestive gland of venus's flytrap. Amer. J. Bot. **55**, 649—657.

SCANDALIOS, J. G., 1969: Genetic control of multiples molecular forms of enzymes in plants: a review. Biochemical Genetics **3** (1), 37—79.

SCHAFFNER, G., 1974: Funktion der Lipasen und der Lipidgranula in der Bäckerhefe. Thesis No. 5287 ETH, Zürich.

SCHEIRER, D. C., 1973: Hydrolyzed walls in the water-conducting cells of *Dendroligotrichum* (Bryophyta): Histochemistry and ultrastructure. Planta **115**, 37—46.

SCHNARRENBERGER, C., A. OESER, and N. E. TOLBERT, 1972: Isolation of protein bodies on sucrose gradients. Planta **104**, 185—194.

SCHNEIDER, E. F., 1972: The rest period of *Rhododendron* flower buds. III. Cytological studies on the accumulation and breakdown of protein bodies and amyloplasts during flower development. J. Exp. Botany **23**, 1021—1038.

SCHNEPF, E., 1963: Zur Cytologie und Physiologie pflanzlicher Drüsen. 3. Cytologische Veränderungen in den Drüsen von *Drosophyllum* während der Verdauung. Planta **59**, 351—379.

— und W. KOCH, 1966: Golgi-Apparat und Wasserausscheidung bei Glaucocystis. Z. Pflanzenphysiol. **55**, 97—109.

SCHULZ, R., and W. A. JENSEN, 1968: Capsella embryogenesis: the egg, zygote, and young embryo. Amer. J. Bot. **55**, 807—819.

SCHUMACHER, W., 1931/32: Über Eiweißumsetzungen in Blütenblättern. Ib. wiss Bot. **75**, 581.

SCHURR, A., and E. YAGIL, 1971: Regulation and characterization of acid and alkaline phosphatase in yeast. J. gen. Microbiol. **65**, 291—303.

SCHWARZENBACH, A. M., 1971: Aleuronevacuolen und Sphärosomen im Endosperm von *Ricinus communis* während der Samenreifung und Keimung. Thesis No. 4645, Swiss Federal Institute of Technology, Zurich.

SCHWENKE, J., G. FARIAS, and M. ROJAS, 1971: The release of extracellular enzymes from yeast by "osmotic shock". Europ. J. Biochem. **21**, 137—143.

SCOTT, W. A., and R. L. METZENBERG, 1970: Location of aryl sulfatase in conidia and young mycelia of *Neurospora crassa*. J. Bacteriol. **104**, 1254—1265.

— K. D. MUNKRES, and R. L. METZENBERG, 1971: A particulate fraction from *Neurospora crassa* exhibiting aryl sulfatase activity. Arch. Biochem. Biophys. **142**, 623—632.

SCRUBB, L. A., A. K. CHAKRAVORTY, and M. SHAW, 1972: Changes in the ribonuclease activity of flax cotyledons following inoculation with flax rust. Plant Physiol. **50**, 73—79.

SEMADENI, E. G., 1967: Enzymatische Charakterisierung der Lysosomenäquivalente (Sphärosomen) von Maiskeimlingen. Planta **72**, 91—118.

SENTANDREU, R., and D. H. NORTHCOTE, 1969: The formation of buds in yeast. J. gen. Microbiol. **55**, 393—398.

SEXTON, R., J. CRONSHAW, and J. L. HALL, 1971: A study of the biochemistry and cytochemical localization of β-glycerophosphatase activity in root tips of maize and pea. Protoplasma **73**, 417—441.

SHAIN, Y., and A. M. MAYER, 1968: Activation of enzymes during germination of a trypsin-like enzyme in lettuce. Phytochemistry **7**, 1491—1498.

SHAW, J. G., 1966: Acid phosphatase from tobacco leaves. Arch. Biochem. Biophys. **117**, 1—9.

SHAW, M., and M. S. MANOCHA, 1965: Fine structure in detached, senescing wheat leaves. Can. J. Bot. **43**, 747—755.

SHELDRAKE, A. R., 1970: Cellulase and cell differentiation in *Acer pseudoplatanus*. Planta **95**, 167—178.

— and G. F. J. MOIR, 1970: A cellulase in *Hevea latex*. Physiologia Plantarum **23**, 267—277.

SHIMODA, C., and N. YANAGISHIMA, 1971: Role of cell wall-degrading enzymes in auxin-induced cell expansion in yeast. Physiol. Plant. **24**, 46—50.

SHUMWAY, L. K., V. CHENG, and C. A. RYAN, 1972: Vacuolar protein in apical and flowerpetal cells. Planta **106**, 279—290.

— J. M. RANCOUR, and C. A. RYAN, 1970: Vacuolar protein bodies in tomato leaf cells and their relationship to storage of chymotrypsin inhibitor I protein. Planta **93**, 1—14.

SIEVERS, A., 1966: Lysosomen-ähnliche Kompartimente in Pflanzenzellen. Naturwiss. **53**, 334—335.

SIMOLA, L. K., 1969: Fine structure of *Bidens radiata* cotyledons, with special reference to formation of protein bodies, spherosomes, and chloroplasts. Ann. Acad. Sci. fenn. **A 4**, 156, 1—18.

SMITH, C. G., 1974: The ultrastructural development of spherosomes and oil bodies in the developing embryo of *Crambe abyssinica*. Planta **119**, 125—142.

SMOLENSKA, G., and S. LEWAK, 1974: The role of lipases in the germination of dormant apple embryos. Planta **116**, 361—370.

SODEK, L., and S. T. C. WRIGHT, 1969: The effect of kinetin on ribonuclease, acid phosphatase, lipase, and esterase levels in detached wheat leaves. Phytochemistry 8, 1629—1640.

SOROKIN, H. P., 1967: The spherosomes and the reserve fat in plant cells. Amer. J. Botany 54, 1008—1016.

— and S. SOROKIN, 1968: Fluctuations in the acid phosphatase activity of spherosomes in guard cells of Campanula persicifolia. J. Histochem. Cytochem. 16, 791—802.

SPICHIGER, J. U., 1969: Isolation und Charakterisierung von Sphärosomen und Glyoxisomen aus Tabakendosperm. Planta 89, 56—75.

SRIVASTAVA, B. I. S., 1968: Increase in chromatin associated nuclease activity of excised barley leaves during senescence and its suppression by kinetin. Biochem. biophys. Res. Comm. 32, 533—538.

— and G. WARE, 1965: The effect of kinetin on nucleic acids and nucleases of excised barley leaves. Plant Physiol. 40, 62—64.

SRIVASTAVA, L. M., and A. P. SINGH, 1972: Certain aspects of xylem differentiation in corn. Can. J. Bot. 50, 1795—1804.

STENLID, G., 1967: On the occurrence of surface-located β-glucosidase and β-galactosidase activity in plant roots. Physiol. Plant. 10, 1001—1008.

STEWARD, F. C., and R. G. S. BIDWELL, 1966: Storage pools and turnover systems in growing and non-growing cells: experiments with C^{14}-sucrose, ^{14}C-glutamine, and ^{14}C-asparagine. J. Exp. Bot. 17, 726—741.

STRAUSS, B. S., 1958: Cell death and "unbalanced growth" in Neurospora. J. gen. Microbiol. 18, 658—669.

SULLIVAN, J. L., and A. G. DEBUSK, 1973: Inositol, less death in Neurospora and cellular ageing. Nature New Biology 243, 72—74.

SUOMALAINEN, H., M. LINKO, and E. OURA, 1960: Changes in the phosphatase activity of baker's yeast during the growth phase and location of the phosphatases in the yeast cell. Biochim. biophys. Acta 37, 482—490.

SUZUKI, T., and S. SATO, 1973: Properties of acid phosphatase in the cell wall of tobacco cells cultured in vitro. Plant Cell Physiol. 14, 585—596.

SWIFT, J. G., and T. P. O'BRIEN, 1972: The fine structure of the wheat scutellum during germination. Austr. J. Biol. Sci. 25, 469—486.

SZE, H., and F. M. ASHTON, 1971: Dipeptidase development in cotyledons of Cucurbita maxima during germination. Phytochemistry 10, 2935—2942.

TAIZ, L., and R. L. JONES, 1970: Gibberellic acid, β-1,3-glucanase and the cell walls of barley aleurone layers. Planta 92, 73—84.

TANIMOTO, E., and Y. MASUDA, 1968: Effect of auxin on cell wall degrading enzymes. Physiol. Plant. 21, 820—826.

TAVENER, R. J. A., and D. L. LAIDMAN, 1972: The induction of lipase activity in the germinating wheat grain. Phytochemistry 11, 989—997.

THOMAS, D. S., M. LUTZAC, and E. MANAVATHU, 1974: Cytochalasin selectively inhibits synthesis of a secretory protein, cellulase, in Achlya. Nature 249, 140—142.

— and J. T. MULLINS, 1967: Role of enzymatic wall softening in plant morphogenesis: Hormonal induction in Achlya. Science 156, 84—85.

THOMAS, P. L., and P. K. ISAAC, 1967: An electron microscope study of intravacuolar bodies in the uredia of wheat stem rust and in hyphae of other fungi. Can. J. Bot. 45, 1473—1478.

THORNTON, R. M., 1968: The fine structure of Phycomyces. I. Autophagic vesicles. J. Ultrastruct. Res. 21, 269—280.

TKACZ, J. S., and J. O. LAMPEN, 1973: Surface distribution of invertase on growing Saccharomyces cells. J. Bacteriol. 113, 1073—1075.

TOMANAGA, G., H. OHAMA, and T. YANAGITA, 1964: Effect of sulphur compounds on the protease formation by Aspergillus niger. J. Gen. Appl. Microbiol. 10, 373—386.

TOMBS, M. P., 1967: Protein bodies of the soybean. Plant Physiol. 42, 797—813.

TREFFRY, T., S. KLEIN, and M. ABRAHAMSEN, 1967: Studies of fine structural and biochemical changes in cotyledons of germinating soybeans. Aust. J. Biol. Sci. 20, 859—868.

TREVITHIK, J. R., and R. L. METZENBERG, 1960: Molecular sieving by *Neurospora* cell walls during secretion of invertase isozymes. J. Bacteriol. **92**, 1010—1015.

TREWAVAS, A., 1972 a: Determination of the rates of protein synthesis and degradation in *Lemna minor*. Plant Physiol. **49**, 40—46.

— 1972 b: Control of the protein turnover rates in *Lemna minor*. Plant Physiol. **49**, 47—51.

TRONIER, B., R. L. ORY, and K. W. HENNINGSEN, 1971: Characterization of the fine structure and proteins from barley protein bodies. Phytochemistry **10**, 1207—1211.

TRUCHET, G., et PH. COULOMB, 1973: Mise en évidence et évolution du système phytolyso-somal dans les cellules des différentes zones de nodules radiculaires de pois (*Pisum sativum* L.). Notion d'heterophagie. J. Ultrastruct. Res. **43**, 36—57.

UEDA, K., 1966: Fine structure of *Chlorogonium elongatum* with special reference to vacuole development. Cytologia Tokyo **31**, 461—472.

VALDOVINOS, J. G., and T. E. JENSEN, 1968: Fine structure of abscission zones. II. Cell-wall changes in abscising pedicels of tobacco and tomato flowers. Planta **83**, 295—302.

— — and L. M. SICKO, 1972: Fine structure of abscission zones. IV. Effect of ethylene on the ultrastructure of abscission cells of tobacco flower pedicles. Planta **102**, 324—333.

VAN DER EB, A. A., and P. J. NIEUDORF, 1967: Electron microscopic structure of the aleurone cells of barley during germination. Acta Bot. Neerland. **15**, 690—699.

VAN DER WILDEN, W., PH. MATILE, M. SCHELLENBERG, J. MEYER, and A. WIEMKEN, 1973: Vacuolar membranes: isolation from yeast cells. Z. Naturforsch. Sect. C **28**, 416—421.

VAN RIJN, H. J. M., 1974: Studies on the biosynthesis of the acid phosphatase, a glycoprotein in baker's yeast (*Saccharomyces cerevisiae*). Thesis University of Utrecht.

— P. BOER, and E. P. STEYN-PARVÉ, 1972: Biosynthesis of acid phosphatase of baker's yeast. Factors influencing its production by protoplasts and characterization of the secreted enzyme. Biochim. biophys. Acta **268**, 431—441.

VARNER, J. E., and R. M. MENSE, 1972: Characteristics of the process of enzyme release from secretory plant cells. Plant Physiol. **49**, 187—189.

— and G. SCHIOLOVSKY, 1963: Intracellular distribution of protein in pea cotyledons. Plant Physiol. **38**, 139—144.

VIGIL, E. L., 1970: Cytochemical and developmental changes in microbodies (glyoxysomes) and related organelles of castor bean endosperm. J. Cell Biol. **46**, 435—454.

— and M. RUDDAT, 1973: Effect of gibberellic acid and actinomycin D on the formation and distribution of rough endoplasmic reticulum in barley aleurone cells. Plant Physiol. **51**, 549—558.

VILLIERS, T. A., 1967: Cytolysosomes in long-dormant plant embryo cells. Nature **214**, 1356—1357.

— 1971: Lysosomal activities of the vacuole in damaged and recovering plant cells. Nature New Biol. **233**, 57—58.

— 1972: Cytological studies in dormancy. II. Pathological ageing changes during prolonged dormancy and recovery upon dormancy release. New Phytol. **71**, 145—152.

VINTÉJOUX, C., 1970: Localisation d'une activité phosphatasique acide dans des inclusions cytoplasmiques des feuilles d'hibernacles chez l'*Utricularia neglecta* L. (Lentibulariacées). C. R. Acad. Sci. (Paris) **270**, 3213—3216.

WADA, S., E. TANIMOTO, and Y. MASUDA, 1968: Cell elongation and metabolic turnover of the cell wall as affected by auxin and cell wall degrading enzymes. Plant Cell Physiol. **9**, 369—376.

WAHEED, A., and S. SHALL, 1971: The relationship between carbohydrate content and enzymic activity of yeast invertase. Enzymologia **41**, 291—303.

WALEK-CZERNECKA, A., 1962: Mise en évidence de la phosphatase acide (monophospho-estérase II) dans les sphérosomes des cellules épidermiques des écailles bulbaires d'*Allium cepa*. Acta Soc. Bot. Polon. **31**, 539—543.

— 1965: Histochemical demonstration of some hydrolytic enzymes in the spherosomes of plant cells. Acta Soc. Bot. Polon. **34**, 573—588.

WARDROP, A. B., 1968: Occurrence of structures with lysosome-like function in plant cells. Nature **218**, 978—980.

WATERKEYN, L., 1967: Sur l'existence d'un « stade callosique » présenté par la paroi cellulaire, au cours de la cytocinèse. C. R. Acad. Sci. (Paris) **265**, 1792—1794.

WESSELS, J. G. H., 1966: Control of cell-wall glucan degradation during development in *Schizophyllum commune*. Anton Leeuwenhoek J. Microbiol. **32**, 341—355.

WESTON, P. D., and A. R. POOLE, 1973: Antibodies to enzymes and their uses, with specific reference to cathepsin D and other lysosomal enzymes. In: Lysosomes in biology and pathology (DINGLE, J. T., ed.), Vol. 3, pp. 425—464. Amsterdam-London-New York: North-Holland/American Elsevier Publishing Companies.

WHEELER, G. E., and A. H. ROSE, 1973: Location and properties of an esterase activity in *Saccharomyces cerevisae*. J. gen. Microbiol. **74**, 189—192.

WIEMKEN, A., 1969: Eigenschaften der Hefevacuole. Thesis No. 4340, ETH-Zürich.

— PH. MATILE, and H. MOOR, 1970: Vacuolar dynamics in synchromously budding yeast. Arch. Mikrobiol. **70**, 89—103.

— and P. NURSE, 1973: The vacuole as a compartment of amino acid pools in yeast. Proc. Third Int. Specialized Symp. on Yeasts, Otaniemi/Helsinki, Part II, pp. 331—347.

WIEMKEN-GEHRIG, V., A. WIEMKEN und PH. MATILE, 1974: Mobilisation von Zellwandstoffen in der welkenden Blüte von *Ipomoea tricolor* (Cav.) Planta **115**, 297—307.

WIENER, E., and J. M. ASHWORTH, 1970: The isolation and characterization of lysosomal particles from myxamoebae of the cellular slime mold *Dictyostelium discoideum*. Biochem. J. **118**, 505—512.

WILEY, L., and F. M. ASHTON, 1967: Influence of the embryonic axis on protein hydrolysis in cotyledons of *Cucurbita maxima*. Physiol. Plant. **20**, 688—696.

WILLIAMS, S. G., 1970: The role of phytic acid in the wheat grain. Plant Physiol. **45**, 376—381.

WILLIAMSON, B., 1973: Acid phosphatase and esterase activity in Orchid mycorrhiza. Planta **112**, 149—158.

WILLIAMSON, C. E., 1950: Ethylene, a metabolic product of diseased or injured plants. Phytopathology **40**, 205—208.

WILSON, C. L., 1973: A lysosomal concept for plant pathology. Amer. Rev. Phytopathol. **11**, 247—272.

— D. L. STIERS, and G. G. SMITH, 1970: Fungal lysosomes or spherosomes. Phytopathology **60**, 216—227.

WINKENBACH, F., and PH. MATILE, 1970: Evidence for *de novo* synthesis of an invertase inhibitor protein in senescing petals of *Ipomoea*. Z. Pflanzenphysiol. **63**, 292—295.

WODZICKI, T. J., and C. L. BROWN, 1973: Organization and breakdown of the protoplast during maturation of Pine tracheids. Amer. J. Bot. **60**, 631—640.

WOOD, R. K. S., 1967: Physiological plant pathology. Oxford and Edinburgh: Blackwell Scientific Publications.

WOOLHOUSE, H. W., 1967: The nature of senescence in plants. Symp. Soc. Exper. Biol. **21**, 179—213.

WYEN, N. V., S. ERDEI, J. UDVARDY, G. BAGI, and G. L. FARKAS, 1972: Hormonal control of nuclease level in excised *Avena* leaf tissues. J. Exp. Bot. **23**, 37—44.

YATSU, L. Y., and T. J. JACKS, 1968: Association of lysosomal activity with aleurone grains in plant seeds. Arch. Biochem. Biophys. **124**, 466—471.

— — 1972: Spherosome membranes, half unit membranes. Plant Physiol. **49**, 937—943.

— — and T. P. HENSARLING, 1971: Isolation of spherosomes (oleosomes) from onion, cabbage, and cottonseed tissues. Plant Physiol. **48**, 675—682.

YOMO, H., and K. SRINIVASAN, 1973: Protein breakdown and formation of protease in attached and detached cotyledons of *Phaseolus vulgaris* L. Plant Physiol. **52**, 671—673.

— and J. E. VARNER, 1971: Hormonal control of a secretory tissue. Curr. Top. Develop. Biol. **6**, 111—144.

— — 1973: Control of the formation of amylases and proteases in the cotyledons of germinating peas. Plant Physiol. **51**, 708—713.

ZALOKAR, M., 1969: Intracellular centrifugal separation of organelles in Phycomyces. J. Cell Biol. **41**, 494—509.

ZANDONELLA, P., 1970: Infrastructure du tissu nectarigène floral de *Beta vulgaris* L.: le vacuome et la dégradation du cytoplasme dans les vacuoles. C. R. Acad. Sci. (Paris) **271**, 70—73.

ZAUBERMAN, G., and M. SCHIFFMANN-NADEL, 1972: Pectin methylesterase and polygalacturonase in Avocado fruit at various stages of development. Plant Physiol. **49**, 864—865.

ZIELKE, H. R., and P. FILNER, 1971: Synthesis and turnover of nitrate reductase induced by nitrate in cultured tobacco cells. J. Biol. Chem. **246**, 1772—1779.

Subject Index

Abscisic acid,
  control of protein turnover 140
  control of senescence 143, 144
Abscission,
  secretion of polysaccharidases 56, 146
  cell wall lytic enzymes involved in
    122—123
  hormonal control 144—146
Abscission zones 121, 146
Acer pseudoplatanus,
  external glycosidases in cultured cells 110
  cellulase in xylem differentiation 124
Acetabularia mediterranea, localization of
  hydrolases 26
Achlya ambisexualis, cellulase in branching
  of hyphae 56, 116, 118, 139
Acid phosphatase 4
  in cell walls 10, 12, 17
  cytochemical localization 10, 12, 36, 38,
    41, 81, 133
  secretion 14, 52, 57
  in vacuoles 22, 27
  structural latency 31
  in aleurone bodies 36
  in spherosomes 38
  in microsomes 58
  in plant pathology 132
Ageing, see senescence
Aleurone bodies,
  isolation 34—36
  structure 34
  storage products 34
  hydrolase content 36
  origin and development 46—50, 93—98
  autophagic activity 104
Alkaline phosphatase 4, 30
Alkaline Pyrophosphatase 17
Allium cepa, cell wall lysis in quiescent
  root meristem 124
Aminopeptidase 3, 14, 21, 27, 30, 33, 94
α-Amylase 10, 22, 57, 135—137
Amylopectin-1,6-glucosidase 7
Anemia rotundifolia, vacuolar dynamics in
  guard cells 42

Angiotensin, hydrolysis by yeast proteases 33
Apical tips of fungal hyphae,
  ultrastructure 52, 115—116
  bursting tendency 112
Apical vesicles in filamentous fungi 52, 115
Apple,
  hemicelluloses during ripening 121
  lipase in cultured embryos 104, 138
Arachis hypogaea, see peanut
Aryl sulfatase 5, 7
Aspergillus nidulans, induction of protease
  production 149
Aspergillus niger, degradation of polygalacturonic acid 128
Aspergillus oryzae,
  β-glucosidase in hyphal tips 112
  protease secretion 149
ATPase 30
Autolysis 74, 90—93, 137, 149
Autophagy 74, 140
  mechanisms of 76—84
  in cellular differentiation and senescence
    84—86
Auxin,
  regulation of growth 139
  control of senescence 141
Avena,
  protein breakdown in detached leaves 74
  cell wall metabolism in coleoptiles 108,
    110, 139
Avocado, softening of fruits 121

Baker's yeast, see *Saccharomyces cerevisiae*
Barley, endosperm, aleurone layer of
  secretion of hydrolases 10, 135—138
  cell wall lysis 10
  sedimentable hydrolases 22, 57
  secretory vesicles 56
  mobilization of storage proteins 48
  aleurone bodies 36
Bean,
  spherosomes in cotyledons 104
  growth of hypocotyl 109
  cellulase in abscission zone 123, 144